DEFENSIVE
MECHANISMS
IN
SOCIAL INSECTS

Edited by

Henry R. Hermann

PRAEGER SPECIAL STUDIES • PRAEGER SCIENTIFIC

New York • Philadelphia • Eastbourne, UK
Toronto • Hong Kong • Tokyo • Sydney

Library of Congress Cataloging in Publication Data

Main entry under title:

Defensive mechanisms in social insects.

Includes bibliographical references and index.
1. Insect societies.   2. Animal defenses.   I. Hermann, Henry R.
QL496.D44   1984   595.7'057   83-24798
ISBN 0-03-057002-6 (alk. paper)

Published in 1984 by Praeger Publishers
CBS Educational and Professional Publishing
a Division of CBS Inc.
521 Fifth Avenue, New York, NY 10175 USA

© 1984 by Praeger Publishers

456789 052 987654321

Printed in the United States of America on acid-free paper

*To Murray Sheldon Blum,
friend, colleague, teacher,
to whom I owe so much*

List of Contributors

Roger D. Akre, Department of Entomology, Washington State University, Pullman, Washington 99164

Alfred Buschinger, Institut fur Zoologie, Techniche Hochchile Darmstadt, 6100 Darmstadt, Federal Republic of Germany

Henry R. Hermann, Department of Entomology, University of Georgia, Athens, Georgia 30602

Ulrich Maschwitz, Fachbereich Biologie der Universitat, D-60000, Frankfurt am Main, Federal Republic of Germany

Andre Quennedy, Universite de Dijon, Faculte des Sciences, Laboratoire de Zoologie, 21100 Dijon, France

Hal C. Reed, Department of Entomology, Washington State University, Pullman, Washington 99164

Stefano Turillazzi, Instituto di Zoologia, Universita di Firenze, Florence, Italy

Preface

Defensive mechanisms in insects are most exquisitely expressed in social species. Nest defense in solitary insects is poorly expressed or lacking, and as the degree of sociality increases, so does the complexity of defensive expression. It is therefore evident that being social and being defensive are by necessity interdependent.

Defensiveness, therefore, stands as an important requirement of sociality. Yet, prior to this volume, defensive mechanisms in social insects have most often been found scattered in the literature, representing bits and pieces of various reports that deal primarily with other broader subjects. Very few reports have been devoted to defensive behaviors in their own light.

It is hoped that this volume will provide some stimulation for future work on defensive mechanisms in the social Insecta, be they morphological or behavioral. Many anatomical structures, their function and the behavior peculiar to their use have been described here as a collective approach to insect defense.

Contents

1
Defensive Mechanisms: General Considerations
Henry R. Hermann

INTRODUCTION

Defensive behavior in any animal species is a behavioral display that offers protection or potential protection to an individual or an individual's group (modified from Hermann and Blum, 1981). The term group here is used to signify any gathering of conspecific individuals. Grouping by conspecific individuals is a characteristic possessed by many animal species and is particularly well expressed in eusocial organisms; this definition, therefore, largely reflects the current and most accepted definitions of sociality.

Wilson most recently (1975) defined society in the Animalia as "a group of individuals belonging to the same species and organized in a cooperative manner." While social groups have a well-defined cooperative nature (social collaboration, West Eberhard, 1981), they also exhibit certain intraspecific competitive behaviors (social competition, West Eberhard, 1981), leading Wynn-Edwards (1962) to define society in terms of "an organization capable of providing conventional competition." The latter concept largely concerns the establishment of social rules and rank within the subgroups of a species, especially in vertebrates, whereas the former concept of cooperative behavior reflects the importance of defensive mechanisms as an adaptive advantage to societies. Other concepts of sociality (Emerson, 1959; Richards, 1965; Richards and Richards, 1951; Wheeler, 1928; Wilson, 1971) impart views that partially or totally are expressed in the definitions presented by Wilson (1975) and Wynn-Edwards (1962).

1

Social insects have four principal classes of natural enemies (as reflected by Starr, 1981): (1) social insect predators, (2) parasites, (3) insect brood parasitoids, and (4) vertebrate predators. The fourth class receives special attention here. In addition, general vertebrate intruders not included in these classes constitute individuals that are looked on by defending organisms in the same light as predators.

Who Defends?

Most animals individually defend themselves to some degree no matter what their stage of social development. There are many defensive strategies within the solitary and social Animalia, from passive exhibits of cryptic, mimetic, and warning coloration (primary defense, Edmunds, 1974) to a diverse assortment of overt behaviors (secondary defense, Edmunds, 1974) (Chadab-Crepet and Rettenmeyer, 1982; Deligne et al., 1981; Hermann and Blum, 1981; Maschwitz and Kloft, 1971; Wilson, 1975).

Many animals that must routinely exhibit defensive behaviors are found in groups or aggregations (see Eickwort, 1981; Michener, 1958). Aggregations may come about because of a variety of reasons, some of which are not necessarily associated with sociality (Hermann, 1979). However, there are obvious survival benefits associated with gregarious behavior (Alcock, 1975; Batra, 1978; Brown and Orians, 1970; Eickwort, 1981; Eickwort et al., 1977; Miller and Kurczewski, 1973; Myers and Loveless, 1976). Buskirk (1981) and Eickwort (1981) point out a number of disadvantages of aggregating. However, mutual defense against nest intruders is of special importance as a positive influence on the phylogeny of social behavior. The reader is also directed to the reports of Allee (1931), Allee and Wilder (1939), Allee et al. (1950), and Gerstell (1939) and the brief reviews by Hermann and Blum (1981) and Wilson (1975) in which examples of group defense are given.

The formation of aggregations and increased density within these aggregations (up to an optimum) often provide a lengthened survival among animals (Allee et al., 1950). Aggregation of conspecific individuals, as pointed out above, is widespread in the animal kingdom, Lin and Michener (1972) and many biologists believing that such behavior may sometimes be antecedent to joining and the formation of semisocial behavior (see review by Eickwort, 1981).

Wilson (1975) supports aggregation, along with its associated sexual behavior and territoriality, as a form of social behavior, defining eusociality within the Insecta (Wilson, 1971) as the culmination of an evolutionary process in which a eusocial insect species will demonstrate: (1) care of the young, (2) a reproductive division of labor, and (3) an overlap in generations. The pathways by which a eusocial existence has been reached apparently differ among the eusocial insect species, passing through a semisocial or subsocial state (Lin and Michener, 1972). Whatever their pathways, however, those species that have reached the eusocial level demonstrate mechanisms of defense in a more pronounced fashion than species at any simpler social level.

Why Defend?

Self-defense is an understandable survival strategy in a solitary species and in a social individual away from its nest. To understand why a species will or will not defend its colony we may use a modification of the cost/benefit scheme recently employed by Emlen (1981) in a discussion of conditions that are worth defending. For instance, if colony investment is low (as in a newly founded colony) the individuals on the nest are relatively nondefensive. As the investment increases, so does the defendability. If the investment is stable (e.g., low parasitism and good environmental conditions), defensive behavior would be more prominent than if it were not stable (there are periods during colony life, however, when parasitism appears to influence increased agitation among colony members, personal observation).

Competition for a single resource very seldom appears to enter into interspecific defense systems in social insects except in the formicids (see section on territoriality in this chapter). Lastly, the type of nest inhabitants and the defendability of a nest based on what the queen does are possibilities of factors that may influence nest defendability but are little known. Some of these factors are considered by Smith (1978) in his analysis of asymmetrics in contests between individuals and the description of bourgeois tactics.

Organized group defense is largely facilitated by altruism (Wilson, 1975), which in turn is dependent on the presence of reproductively neuter workers that may go so far as to give up their life for the protection of the colony (as in sting autotomous species, Hermann, 1971; Maschwitz and Kloft, 1971; Pursley, 1973; Sakagami and Akahira, 1960). Altruism also is considered to be a phenomenon of such importance that it forms the basis for insect sociality (Craig, 1979; Crozier, 1979) in which an individual's reproductive performance is reduced in order to increase the reproductive performance of other individuals. However, unless the cost–benefit ratio is low enough for altruism to arise by kin selection, selection will favor daughters capable of resisting maternal manipulation (Craig, 1979; Rubenstein and Wrangham, 1980). Parental manipulation (Alexander, 1974; Michener and Brothers, 1974) accounts for the evolution of altruism by selection on parents while the kin selection theory (Hamilton, 1963, 1964) "is based on the postulate that the decreased fitness of altruists can be more than compensated for by the increased fitness of relatives benefiting from the altruism" (Craig, 1979).

How defensive mechanisms themselves have evolved to be best expressed in well-defined defenders of polymorphic colonies of the eusocial Insecta probably depends on an optimum trade-off between the colony's investment in a soldier caste and a worker caste (Deligne *et al.*, 1981; Oster and Wilson, 1978). Investment in a soldier caste biotechnologically leads to: (1) a decrease in size of the foraging force and (2) an increase in potential colony survival: thus, a balance between these castes apparently must be reached as a function of colony fitness.

SPECIES THAT DEFEND

All animals that form groups defend themselves, their group, or their nests in some way. As pointed out briefly earlier in this chapter, substantial evidence exists to show that there is great effectiveness in group defense. Group defense is selected for if its positive effects on reproductive success outweigh the negative effects of proximity to competitors (West Eberhard, 1979). Wilson cites examples among the arthropods in tortoiseshell and peacock butterflies (Mosebach-Pukowski, 1937), *Ascaloptynx* larvae (Henry, 1972) and juvenile spiny lobsters, spider crabs, and king crabs (Powell and Nickerson, 1965; Stevcic, 1971), all of which benefit defensively because of their aggregating habits. Other examples may be found in the report by Hermann and Blum (1981). Eusocial insects are characteristically grouped, and as pointed out earlier it is in this category of the Insecta that we find the best defined examples of defensive behavior.

Eusociality actually is characteristic of relatively few members of the Insecta, having arisen 11 or more times in the order Hymenoptera and once in the Isoptera. However, a number of insect groups that are not eusocial (included in the following list) do possess some social traits.

Isoptera

According to Krishna and Weesner (1969, 1970), the 1900 known living and fossil termite species, all of which are eusocial, represent six families: The Mastotermitidae, Kalotermitidae, Hodo-termitidae, Rhinotermitidae, Serritermitidae, and Termitidae. Families other than the Termitidae are considered the lower termites. Other accounts of taxonomic significance are mentioned in Krishna and Weesner's two-volume treatise on termites.

Deligne *et al.* (1981) and Stuart (1969) have given general accounts of defensive behavior in termites, pointing out that even though ants have been considered the primary enemies of termites (Bugnion, 1922; Wheeler, 1936), vertebrates are also extremely important in this respect (Hegh, 1922). Bouillon (1970) and Moore (1969) have also reviewed forms of defense among termites.

Hymenoptera

Sphecoidea

Sphecidae (Pemphredoninae). Only species of the genus *Microstigmus* of the family Sphecidae have attained eusociality, and social behavior in this genus is best known in *Microstigmus comes* (Matthews, 1968). Other species of this genus may later prove to be eusocial (Matthews, 1970; also see Eickwort, 1981). Although the defensive behavior of this species has not been described in detail, the species does utilize a form of architectural defense by building a nest suspended by a long pedicel. Both males and females cooperate in nest defense by responding quickly to nest disturbances. Up to 11 females apparently cooperate in building and guarding their single communal nest. Whether smearing of a repellent substance on the pedicel occurs as it does in the Vespidae (Hermann and Dirks, 1974; Jeanne, 1970) is not known. According to Matthews (personal communication), complex defensive behaviors are not apparent in colonies of this species. Nest location may be its most important means of defense. *Trigonopsis cameronii* (Eberhard, 1972; West Eberhard, 1981) lives a communal existence, cooperating in constructing a nest, but it lacks some other qualities to make it eusocial.

Vespoidea

Eumenidae. Most species of this cosmopolitan family are solitary and a few species are subsocial (Evans, 1958; Evans and West Eberhard, 1970).

Vespidae. This family consists of 800–1000 species comprising three social subfamilies, the Stenogastrinae, Polistinae, and Vespinae

(Edery *et al.*, 1978; Snelling, 1981), and four nonsocial subfamilies. The Stenogastrinae are confined to Southeast Asia and apparently possesss some social traits (Edery *et al.*, 1978; Iwata, 1967; Williams, 1919).

The Polistinae consists of three tribes: the Ropalidiini, with diverse nesting behavior (Yoshikawa, 1964); the Polybiini, with a wide variety of social behavior (Edery *et al.*, 1978; Jeanne, 1975; Schremmer, 1973); and the Polistini, with open, single-comb nests.

Vespidae (Stenogstrinae) are subsocial (Iwata, 1967; Yoshikawa *et al.*, 1969). This subfamily includes six or seven Old World genera, consisting of approximately 50 species (Snelling, 1981).

Vespidae (Polistinae) Ropalidiini–*Ropalidia* (from the Old World) is quasisocial (Iwata, 1969; Spradbery, 1973; Yoshikawa, 1964).

Vespidae (Polistinae) Polybiini are eusocial (Evans and West Eberhard, 1970; Jeanne, 1972; Pardi and Marino Piccioli, 1970; Richards, 1969; Richards and Richards, 1951; Schremmer, 1972; Spradbery, 1973; Wilson, 1971).

Vespidae (Polistinae) Polistini are considered to be primitively eusocial (Deleurance, 1957; West Eberhard, 1969; Guiglia, 1972; Hermann and Dirks, 1975; Pardi, 1948; Spradbery, 1973; Yamane, 1971; Yoshikwa, 1963).

Vespidae (Vespinae), with approximately 80 species (Snelling, 1981), are advanced eusocial (Guiglia, 1972; Ishay and Landau, 1972; Ishay *et al.*, 1967; Kemper and Dohring, 1967; Spradbery, 1973).

This last subfamily, Vespinae, consists of three major eusocial genera: *Provespa, Vespula,* and *Vespa* (see Snelling, 1981, for discussion of other recognized genera). *Provespa* species (three) are restricted to Borneo, Malaya, and Sumatra. The *Vespula* (about 30 species) are primarily Holarctic in distribution, except for two species that have expanded southward (Edery *et al.*, 1978). Species of the subgenus *Dolichovespula* usually build aerial nests (Yamane, 1975), while other species of the genus *Vespula* build mainly subterranean nests. A few species of the latter genus are known to build aerial nests (Weyrauch, 1935: Thomas, 1960). Edery *et al.* (1978) give an account of the distribution of the remaining *Vespula* species. Species of the genus *Vespa* mainly occur in the Indo-Australian archipelago.

Formicoidea

Formicidae. All members of this family (between 12,000 and 15,000 species) (Snelling, 1981) are strongly eusocial. Ten subfamilies are currently recognized: Sphecomyrminae (extinct), Myrmeciinae, Ponerinae, Leptanillinae, Dorylinae, Ecitoninae, Pseudomyrmecinae,

Myrmicinae, Dolichoderinae, and Formicinae (Wilson, 1975) (also see Snelling, 1981; Taylor, 1978).

Apoidea

Halictidae. Members of this group may be solitary, communal, parasocial, quasisocial, semisocial, or primitively eusocial (Halictinae: Halictini and Augochlorini; Nominae and Dufoureinae) (Michener and Brothers, 1974; Snelling, 1981).

Andrenidae. Most species in this family are solitary but a few are communal (Michener and Brothers, 1974).

Megachilidae. Most species of this family are communal or quasisocial (Michener and Brothers, 1974).

Anthophoridae. Members of this group are largely colonial or communal (Michener and Brothers, 1974). The allodapines (Xylocopinae: Ceratinini) are subsocial or primitively eusocial.

Apidae (Bombinae). Members of this group are either solitary, communal, quasisocial, or primitively eusocial (Dodson, 1966; Free and Butler, 1959; Michener and Brothers, 1974; Roberts and Dodson, 1967; Sakagami and Zucchi, 1965; Sladen, 1912; Zucchi *et al.*, 1969).

Apidae (Apinae), including the tropical Meliponini and Apini, are highly social.

The families Colletidae, Oxaeidae, Melittidae, and Fedeliidae, listed by Snelling (1981), do not include social species.

FORMS OF DEFENSIVE BEHAVIOR

A detailed discussion on avoidance and attack behaviors may be found in the report by Hermann and Blum (1981) and the following chapters of this book. A brief review of these and some additional defensive behaviors found in the Isoptera and eusocial Hymenoptera follows.

Individual Defensive Behavior

At the individual level within the class Insecta, defensive behavior is expressed in such a way that solitary and social species behave alike

(Hermann and Blum, 1981; Spradbery, 1973). Females of solitary insect species (Eickwort, 1981) have no colonies (and hence no overlap in generations, no division of labor, and no queen ruling the nest), construct no well-defined nest, and have no large investment to protect. Their investment (brood) is most often protected to varying degrees by the manner and location in which the brood chamber is constructed (Evans and West Eberhard, 1970). Further behavioral protection after nest closure is lacking in solitary forms (except in certain cases, Steiner, 1975) whereas the characteristics of eusociality warrant continuous colony protection, especially when the investment becomes large.

With varying degrees of sociality and presociality found among the diverse Insecta there are varying degrees of cooperation in the nesting areas (Eickwort, 1981), varying degrees of investment (Starr, 1979), and therefore, varying degrees of defense.

Major Types of Nest Defense

Escape and Other Avoidance Behaviors

Individual defense may be nonaggressive, in which case the insect may seek protection by escaping an encounter with an intruder. Seeking shelter, remaining on the back of the nest, taking flight, or leaving the nest in some way, dropping, and defensive immobility are examples of nonaggressive defense (Akre et al., 1976; Deligne et al., 1981; West Eberhard, 1969; Edery et al., 1978; Green et al., 1976; Hermann and Blum, 1981; Hermann and Dirks, 1974; Hermann and Young, 1980; Jeanne, 1970; Landolt and Akre, 1979; Landolt et al., 1977; Lecomte, 1951; Maier and Maier, 1970; Pliner et al., 1975; Ribbands, 1964; Turillazzi, 1979; Turillazzi and Ugolini, 1978).

Escape is also demonstrated on and in the vicinity of the nest of a social species, depending on the species, age of the nest occupants, and degree of investment. Escape, therefore, is an avoidance reaction, enabling the individual to protect or potentially protect itself from danger. Such behavior is advantageous to an individual of a solitary species in which there is no investment to protect, and it is advantageous to an individual of a social species but disadvantageous to its colony if the individual's colony is at stake. Escape by colony members at the expense of losing the queen and brood can be disastrous to the colony.

Other forms of avoidance in which the nesting area is protected in various ways are common among social insect species. Examples are: (1) seeking sheltered nesting locations (e.g., in cavities) Iwata,

1971; Yamane, 1976; Yamane and Makino, 1977), (2) building a sheltered nest (e.g., closed nests of vespine wasps, some polybiine wasps, and termites) (Deligne, Quennedey, and Blum, 1981; Evans and West Eberhard, 1970; Grasse and Noirot, 1948; Jeanne, 1975; Noirot, 1970; Wood and Sands, 1978; Yamane and Makino, 1977), (3) other architectural modifications (e.g., nest pedicels, entrance turrets, and entrance blockage) (Gentry, 1974; Hermann and Dirks, 1974; Jeanne, 1970, 1972, 1977, 1979; Matthews, 1968; Michener, 1962; Sakagami and Michener, 1962; Stephen *et al.*, 1969), (4) living symbiotically with other organisms that offer protection (e.g., ant–wasp associations) (Evans and West Eberhard, 1970; Hamilton, 1972; Richards and Richards, 1951), (5) using secretions to smear or otherwise put on or near the nest as repellents (West Eberhard, 1969; Edery *et al.*, 1978; Green *et al.*, 1976; Hermann and Dirks, 1974; Jeanne, 1970, 1977; Kerr and de Lello, 1962; Landolt and Arke, 1979; Landolt *et al.*, 1977; Turillazzi, 1979; Turillazzi and Ugolini, 1978), (6) using sticky fluids as entrapping devices (Drory, 1872; Kerr and de Lello, 1962; Maidl, 1934; Maschwitz and Maschwitz, 1974; Stephen *et al.*, 1969) and (7) use of aposematic colors as a warning device. However, social insects most often react to a nest intruder in an overtly antagonistic manner. Overt defense is expressed as either warning behavior or attack.

Warning Behaviors

Social Insects Demonstrating Warning Behaviors. Defensive behavior in animals in general usually includes, at least in part, threatening or bluffing. Threats and bluffs are categorized here as warning behaviors, behaviors that symbolically tell an intruder to stay away from the nest. Conspicuous alert reactions by disturbed members of *Polistes* colonies have been reported often (Corn, 1972; Hermann and Blum, 1981; Hermann *et al.*, 1975; Rau, 1930; Schmitt, 1920; Spieth, 1948; Starr, 1981; West Eberhard, 1969; Weyrauch, 1928; Yoshikawa, 1963).

Warning behaviors have been shown to be well developed in vespid wasps (Hermann and Blum, 1981), particularly in polistine species, which have an exposed nest and therefore are in direct contact with any intruder that approaches the nest. They are known to hang from the nest by their meso- and metathoracic legs and wave their prothoracic legs, pump their abdomen, lunge forward, raise their wings, extend their antennae, and turn their sting toward an intruder (West Eberhard, 1969; Evans and West Eberhard, 1970; Hermann and Blum, 1981; Jeanne, 1980, Starr, 1981).

Benefits of Warning Behaviors. Warning behaviors allow the defending insect to avoid directly contacting an intruder and possibly losing its life. As pointed out by Tinbergen (1974), it is a distinct disadvantage when individual animals are actually damaged or killed and thus excluded from reproduction. In the case of eusocial Hymenoptera, damage or loss of life represents a comparable disadvantage if such a colony loss affects the reproductivity of the queen or the functions of the colony that facilitate raising of the young.

Warning behaviors do not necessarily culminate in attack, although it has been suggested by West Eberhard (1969) and explicitly stated by Pardi and Marino Piccioli (1970) that a threat display "has the obvious significance of intention movements of an attack flight" (Starr, 1981). In contrast to the reports of Frings and Frings (1975) and Starr (1981) I am not convinced that "threats arise as intention movements of combat." Warning behaviors often can help a defender avoid further defensive function. If the intruder recognizes the warning behavior and subsequently refrains from further intrusion, attack will most likely never come about. Discontinuous intrusion, in which the intruder progresses toward a nest to a point of eliciting warning behavior but stops before attack ensues, characteristically terminates in habituation on the part of the defender. Continuous intrusion, however, in which warning behaviors do not represent a threat to an intruder that is significant enough to halt intrusion, usually ends in attack.

Warning and Attack Thresholds. The degree of warning and attack exhibited by individuals of a species depends on a threshold level for defense (Fletcher, 1978; Hermann and Blum, 1981). The threshold level, in turn, depends on many things, for example (1) the species, (2) the type of individuals defending the nest, (3) the stage of nest development (investment), and (4) the degree of parasitism present (personal observation). Males and fertile females of polistine species may demonstrate warning behaviors (Starr, 1981), but their threshold for attack is poorly defined or nonexistent. Beyond warning behaviors, males and fertile females generally resort to escaping rather than having an altercation with an intruder. A nest in early development has only a few cells and one or more fertile females (cofoundresses in pleometrotic species), all of which are reluctant to sting. Although the first workers that emerge will defend the nest, they are not as much a threat to an intruder as workers on an older and larger nest. As the nest grows in size the threshold for attack is lowered and strong overt defensive behaviors are demonstrated (Akre *et al.*, 1976; Bristowe, 1932; Butler and Free, 1952; Cott, 1940; Edery *et al.*, 1978;

Fletcher, 1978; Gary, 1974; Gaul, 1953; Green *et al.*, 1976; Hamilton, 1964; Hermann and Blum, 1981; Hermann and Dirks, 1975; Hinde, 1970; Kerr and de Lello, 1962; Kerr and Cruz, 1961; Landolt *et al.*, 1977; Lecomte, 1951; Lorenz, 1974; Pliner *et al.*, 1975; Potter, 1965; Rau and Rau, 1970; Ribbands, 1964; Richards, 1953; Spradbery, 1973; Stephen *et al.*, 1969).

Attack

Types of Attack. Attack is carried out in several ways. Social insects generally demonstrate attacks in which more than one individual leaves the nest together in an attempt to subdue the intruder. Individuals in the groups may leave independently or they may collectively leave the nest due to visual, auditory, or chemical communication (Akre, *et al.*, 1976; Hermann, 1975; Hölldobler, 1977; Ishay, 1973; Lindauer, 1967; Markl, 1965, 1967, 1968, 1970, 1973; Markl and Fuchs, 1972; Sakagami, 1960; Spangler, 1967).

Degrees of readiness to defend their nest (threshold differences as indicated above) are species specific (Akre *et al.*, 1976; Butler and Free, 1952; Green *et al.*, 1976; Hamilton, 1964; Lecomte, 1951; Ribbands, 1964; Stephen *et al.*, 1969). Hermann and Blum (1981) and Sakagami (1976) review some of the differences in selected social Hymenoptera. Fletcher (1978) points out differences in aggressiveness even between races of honey bees due to temperature and geographic variation.

Behavioral and Morphological Defensive Mechanisms

Eusocial insect species may employ a behavioral or morphological defense, and usually both types are involved. Morphological defense usually involves either the venom apparatus or the mandibles. The venom apparatus is discussed in Chapter 6 of this volume. One or more of the behavioral defense mechanisms in the following list may be found in every group of social insects.

Passive and Overt Behaviors

Abdominal Bursting

This is found in certain formicid species of the genus *Camponotus* (Formicinal). Hydrostatic pressure within the gaster causes the mandibular glands and the gaster that houses them to burst, releasing large quantities of sticky fluid that functions in immobilizing attacking predatory species (Maschwitz and Maschwitz, 1974).

Alert Posture

This has been described by Starr (1981) as a substrate for other behaviors of a defensive nature in polistine species.

Antennate Air

Antennate air may function in polistine wasps purely as a sensory movement (Starr, 1981), serving to increase information on potential danger by chemoreception, although aposematic coloration may facilitate an overall picture received by a nest intruder.

Ant–Wasp Symbioses

Certain members of vespid species in the tribe Polybiini live symbiotically with formicid species in the subfamily Dolichoderinae, each wasp species possibly living in mutualism with the other (Evans and West Eberhard, 1970; Hamilton, 1972; Jeanne, 1978; Rettenmeyer, 1963; Richards and Richards, 1951). Intratribal benefits may be acquired from this relationship in addition to interfamilial benefits.

Aposematic Coloration

Found in some bees and wasps, aposematic coloration appears to play a role of support in warning behaviors (Evans and West Eberhard, 1970; Hermann and Blum, 1981). Male social wasps are known to threaten an intruder (Kemper and Dohring, 1967; Little, 1979; Marino, Piccioli and Pardi, 1970; Steiner, 1932; Starr, 1981). Mimicking insects often overdisplay their aposematic colors (personal observation) and it is my belief that this is what may be occurring in some male wasps. Males also display warning behaviors such as leg waving, gaster bending, wing fluttering, wing flipping, and wing buzzing (Starr, 1981), and realistic stinging motions, forms of automimicry (Brower *et al.*, 1967).

Architectural Defense

A wide variety of behaviors may be found in this category, incorporating: (1) cell walls, caps, multiple combs, and nest envelopes; (2) entrance turrets; (3) nest camouflage: (4) nest entrance blockage; (5) construction of nest pedicels (Akre *et al.*, 1981; Evans and West Eberhard, 1970; Friese, 1914; Gentry, 1974; Hermann and Blum, 1981; Hermann and Dirks, 1974; Iwata, 1971; Jeanne, 1970, 1972, 1975, 1977, 1979a, 1979b; Kerr and de Lello, 1962; Michener, 1962; Rau and Rau, 1970; Rodriguez, 1968; Sakagami and Michener, 1962; Stephen *et al.*, 1969; Yemane and Makino, 1977), and (6) construction of multiple combs (Gibo, 1977, 1978; Jeanne, 1979a; Strassmann, 1981); (7) construction of inner and outer termite nest walls, sometimes separated by an air space (Deligne, Quennedey, and Blum, 1981; Grasse and Noirot, 1948; Noirot, 1970); (8) division of termite nests into chambers that are connected only by small openings (Deligne, Quennedey, and Blum, 1981); (9) storing of sealing materials within the termite nest (Deligne, Quennedey, and Blum, 1981); and (10) building of galleries through which termites forage (Wood and Sands, 1978).

Biting and Spraying of Venom

These behaviors are demonstrated by many hymenopterous species but primarily by doryline, myrmicine, and formicine ant species, stingless bees, and certain vespid wasps (Brown, 1976; Drory, 1872; Gotwald, 1974; Green *et al.*, 1976; Hermann and Blum, 1968; Hermann *et al.*, 1975; Hung and Brown, 1964; Kerr and Cruz, 1961; Kerr and de Lello, 1962; Maschwitz, 1964; Maschwitz and Kloft, 1971; Saslavasky *et al.*, 1973; Stephen *et al.*, 1969). Biting is also demonstrated in several diverse ways by termites (Deligne, Quennedey, and Blum, 1981), including simple biting, reaping, symmetric snapping, and asymmetric snapping by soldiers. In addition to biting, mandibular and salivary secretions may facilitate the effects of biting by termites.

Biting and Stinging of Foreign Plants and Animals

These behaviors are known to be demonstrated by pseudomyrmecine and dolichoderine ant species (Janzen, 1967).

Mandibular Clicking

The snapping together of mandibles is demonstrated by ant species in the genera *Acanthognathus, Anochetus, Odotomachus, Strumigenys,* and *Daceton* (Brown, 1976; Brown and Wilson, 1959; Mayr, 1892, 1893; Wroughton, 1892). Symmetric and asymmetric snapping has been described for species of *Tuberculitermes, Termes, Capritermes,* and *Machadotermes* by Deligne, Quennedey, and Blum (1981).

Chemical Alarm

Chemical alarm is well known in ants (Blum and Hermann, 1978a, 1978b). It is also known for some vespine wasps (Edwards, 1980), but it is little known in polistine wasps (Jeanne, 1972; Rau, 1931; Starr, 1981).

Darting

Darting is a lunging forward, sometimes in a rhythmic fashion, on or off the nest (Owen, 1962; Starr, 1981; West Eberhard, 1969) in a polistine wasp colony.

Defensive Immobility

Defensive immobility is a behavior found associated with escape in many hymenopterous species (Akre *et al.*, 1976; West Eberhard, 1969; Edery *et al.*, 1978; Green *et al.*, 1976; Hermann and Dirks, 1974; Hermann and Young, 1980; Jeanne, 1970; Landolt and Akre, 1979; Landolt *et al.*, 1977; Lecomte, 1951; Maier and Maier, 1970; Pliner *et al.*, 1975; Ribbands, 1964; Turillazzi, 1979; Turillazzi and Ugolini, 1978; Wheeler, 1910).

Defensive Smearing

Defensive smearing involves the smearing of repellent compounds on the nest and nest pedicel by polistine and polybiine wasps. The repellent chemicals are produced from glands on abdominal sterna in females (Delfino *et al.*, 1979; Hermann and Dirks, 1974; Landolt and Akre, 1979; Jeanne, 1970, 1977; Post and Jeanne, 1980; Turillazzi, 1979; Turillazzi and Ugolini, 1978). Glands opening at the base of the fifth gastral sternite produce a substance smeared on leaves and other types of substrate that functions as a trail pheromone during nest evacuation and subsequent emigration (Jeanne, 1981).

Escape

Escape has been briefly reviewed earlier in this chapter. *See also* Retreat.

Forward Jerking and Thrusting

These behaviors are known to be demonstrated by certain polistine wasps (West Eberhard, 1969; Evans and West Eberhard, 1970; Hermann and Blum, 1981). *See also* Darting.

Frontal Weapons

Some termites have well-defined frontal weapons composed of frontal glands, the frontal pore and cephalic structures from the frons, clypeus, and labrum that function in a form of chemical defense (Deligne *et al.*, 1981). Frontal modifications consist only of local frontal differentiation or formation of a frontal gutter or frontal pore.

Gaster Bending

In this phenomenon, the gaster is arched, the distal end pointing toward the substrate or to one side (Starr, 1981).

Group Defense

Group defense in the social Insecta has led to the development of an elaborate communication system. Alarm behavior in this class is primarily elicited by alarm pheromones, widely distributed in the social Hymenoptera, that appear to be produced in greater concentrations than any other exocrine gland products.

Although alarm pheromones are known to be produced in vespids (Maschwitz, 1964) and various bees (Blum *et al.*, 1978; Shearer and Boch, 1965), ants have proven to be the richest source of alarm pheromones among the social insects (Hermann and Blum, 1981). Sources of alarm substances have been the mandibles, venom, and Dufour's glands (Blum and Hermann, 1978a, b).

Leg Waving

Leg waving has been seen in certain New World polistine wasp species (West Eberhard, 1969; Evans and West Eberhard, 1970; Hermann and Blum, 1980; Starr, 1981). It was also reported in the African polistine, *Belonogaster griseus* (Marino, Piccioli, and Pardi, 1970).

Mandibular Gland Secretions

Secretions of the mandibular gland are used defensively primarily by ant species, apids and vespids (Buren *et al.*, 1970, 1977; Hermann *et al.*, 1971; Hockings, 1884; Kerr and Cruz, 1961; Kerr and de Lello, 1962; Marianno, 1910; Maschwitz, 1964; Stephen *et al.*, 1969).

Phragmotic Head

Head structures that are used to block nest entrances and an account of false phragmosis have been reported for species of *Cryptocerous* and *Colobopsis* (Brown, 1967; Wilson, 1971, 1974) and *Pheidole* (Buren *et al.*, 1977), respectively. Phragmotic termite soldiers from the family Kalotermitidae (Deligne *et al.*, 1981) are known.

Pumping of Abdomen

This is a behavor found in the Vespidae (Hermann and Blum, 1981).

Raising of Wings

This is also a behavior found in the Vespidae (Hermann and Blum, 1981).

Retreat

As described by Starr (1981), polistine wasps move to the far side of the nest, out of sight.

Sticky Fluids

Using sticky fluids to hinder intrusion has been found in the Bombinae and Meliponidae (Drory, 1872; Kerr and de Lello, 1962; Maidl, 1934; Maschwitz and Maschwitz, 1974; Stephen *et al.*, 1969).

Sting Autotomy

The phenomenon of self-amputation of the sting is called autothysis by Maschwitz and Kloft (1971). It is found to occur in the genera *Apis* (Apidae) and *Pogonomyrmex* (Formicidae) as well as certain polybiine and polistine wasps (Daly, 1953; Darwin, 1859; Hermann, 1971; Janzen, 1967; Maschwitz, 1964; Maschwitz and Kloft, 1971; O'Connor *et al.*, 1963; Pursley, 1973; Rietschel, 1937; Sakagami and Akahira, 1960). Sting autotomy comes about in response to the presence of large lancet barbs.

Stridulation

Production of defensive sounds has been reported in myrmicine and ponerine ant species and in some species of wasps (Akre *et al.*, 1976; Hermann, 1968; Hölldobler, 1977; Ishay, 1973; Lindauer, 1967; Markl, 1965, 1967, 1969; Markl and Fuchs, 1972).

Territoriality

An expansion of our known examples of territoriality within the eusocial Insecta has primarily been due to the recent investigations of Hölldobler (Hölldobler, 1974, 1976a,b, 1979; Hölldobler and Wilson, 1970, 1977a,b). Hölldobler (1978, 1979) defines a territory as an area that an animal or animal society uses exclusively and defends against intra- and interspecific intruders. Reviews of territoriality may be found in the reports of Baroni-Urbani (1979) and Hermann and Blum (1981).

Forms of territorial behavior in eusocial insects have been studied primarily in species of ants. Most cases of territorial disputes in this group apparently are associated with a foraging area or possible overcrowding.

Intraspecific and interspecific tolerance varies among species, large degrees of variation sometimes existing even among members of a single formicid subfamily (Hölldobler, 1974; Weber, 1972). Extreme examples of low and high intersubfamilial tolerances to outsiders are found in *Oecophylla longinoda* (Hölldobler, 1978, 1979; Hölldobler and Wilson, 1977) and members of the tribe Attini (Weber, 1956, 1972), respectively. *Oecophylla longinoda* tolerates almost no other ant species in the trees they inhabit, and they even exclude each other, demonstrating aggressive interactions so severe that narrow, unoccupied areas are, in effect, "no ant's land." On the other hand, Weber (1972) reports that most attine species show little or no hostility when their nests are disturbed. Some colonies even readily fuse with one another (Weber, 1956).

Territoriality is displayed in several known ways. Territorial demarcation (Hermann and Blum, 1981) has been found to occur in *O. longinoda* (Hölldobler, 1978, 1979), some species of *Pogonomyrmex* (Hölldobler, 1974), *Cremtogaster* (Decelles, 1976), *Messor* (Blancheteu, 1975), *Aphenogaster* (Fellers and Fellers, 1976), and dorylines (Gotwald, 1976; Rettenmeyer, 1963). When further investigations are carried out it is certain that the list of species demonstrating territorial behavior in the form of demarcation will become much larger.

Wilson (1971) distinguishes three classes of foraging by various ant species, some of which are distinctly territorial: (1) early arrivers, which avoid most encounters with other foraging species; (2) extirpators, which are distinctly aggressive, entering combat with competitive species; and (3) insinuators, which occupy a food site with an extirpator species, making every attempt to avoid eliciting aggressive responses from them.

Ant species that normally are highly tolerant of noncolony members may be coerced into demonstrating high degrees of territorial behavior. Attines of different species generally stand still when meeting each other during the course of their foraging, slightly spreading their mandibles and antennae. Most attine species show little or no hostility even when their nests are disturbed, spending their energy recovering brood and garden fragments. When one species is introduced into the nest yard of another species, however, aggressive behavior is strongly demonstrated by the territorial defenders (Weber, 1972).

Similar territorial behavior has been reported in *Paraponera clavata* (Hermann and Young, 1980), when conspecific individuals were displaced from their own nest yard to that of another colony. Such displaced ants were attacked and dismembered, and any hostility exhibited by the intruder brought on other colony defenders that assisted in capturing and dismembering the intruder (Mobbing).

The concept of territoriality in investigations of wasps is in an early stage of development. Intraspecific competition among vespines at a food source has been reported by Free (1970). Polistines and polybiines are apparently not active competitors for food.

Warning Displays

Many of the individually listed behaviors are warning or threat displays that have been summarized in the Warning Behavior section of this chapter. Jeanne (1980) reviews some of the warning behaviors of social wasps as alarm mechanisms. Starr (1981) predicts that organized threat displays will generally be found in species with large, strongly stinging individuals and also where nests and habits of defenders are constantly in clear view.

Wing Movements

These have been subcategorized by Starr (1981) as wing-buzz, wing-flip, wing-flutter, and wing-raise.

Additional information on defensive behaviors in the eusocial Hymenoptera may be found in the report by Hermann and Blum (1981). Defensive behavior in termites has been discussed by Deligne *et al*. (1981).

PHYSIOLOGY OF SOCIAL HYMENOPTERAN AGGRESSION

Before discussing the little known field of the physiology of aggression it will be necessary to briefly point out the types of aggression prevalent in social hymenopterous insects.

Types of Aggression

Aggression may be divided into interspecific and intraspecific types. Interspecific aggression deals with relationships between the social species and their predators, general intruders, and parasitoids. Reaction on the part of the social species in this case is certainly what we generally refer to as defensive behavior.

Intraspecific aggression may be of two types: (1) dominance reactions (Strassmann, 1981; personal observations) or (2) intrusion by foreign conspecific wasps. The latter category may sometimes be listed under colony defense; however, dominance reactions, most obvious at the beginning of nest founding, should not be listed as a defensive strategy in the context of this chapter. It is true that there is an obvious connection between aggression and dominance in all animals (Bernstein, 1976), but the relationship between dominance aggression and defensive aggression is unclear.

Center for Aggression

Aggression most certainly must be influenced by endocrine function. In vertebrates, much aggressive behavior can be associated with a specific part of the brain called the lymbic system. Abnormal function of the lymbic system often causes increased or decreased aggression, sometimes with disastrous results.

Aggression in the establishment of dominance hierarchies within the social Hymenoptera has been shown to function in relation to increased amounts of juvenile hormone (JH) produced in the corpora allata of *Polistes annularis* (Barth *et al.*, 1975). Enlarged corpora allata correlate with increased JH production, increased ovary size, and dominant position in the hierarchy. According to Roseler *et al.* (1980), the corpora allata (CA) in all subordinated females of *Polistes gallicus* were found to be smaller and less active than those in dominant females. Barth *et al.* (1975) induced egg maturation in subordinated females of *P. annularis* with topical application of JH. Increased ovary development due to exogenous JH was also shown for *Polistes metricus* (Bohm, 1972).

Thus, oogenesis and dominance aggressiveness have been shown to be controlled by hormones. However, little if any information has been presented on defensive aggressiveness. The queen appears to exhibit more warning behaviors in some cases when the immediate vicinity of the nest is approached by an intruder (personal observation). This may certainly be related to her dominant position. However, whether the queen will leave the nest in defense of her investment is not well understood.

Alpha females of a pleometrotic species generally remain on the nest, leaving the nest chores other than egg laying up to subordinate cofoundresses. To leave the nest in defense of her colony would threaten her personal fitness and negate the benefits of being dominant. The cost/benefit ratio would be too high for such an act to come about.

In light of this brief review, the relationship between defensive and dominance aggressiveness and their association with the corpora allata and level of JH appears to be a lucrative area of research. Konishi (1971) stated that in animals as a whole the central nervous system must often be left as an unknown quantity, a black box in both ethological and neurophysiological studies. This is indeed the case in studies of this nature in relation to aggressive behaviors in the Insecta.

REFERENCES

Akre, R. D., W. B. Garnett, J. F. MacDonald, A. Green, and P. Landolt. (1976). Behavior and colony development of *Vespula pennsylvania* and *V. antropilosa* (Hymenoptera: Vespidae). *J. Kans. Entomol. Soc.* 49, 63–84.

Akre, R. D., A. Green, J. F. MacDonald, P. J. Landolt, and H. G. Davis. (1981). The yellowjackets of America north of Mexico. Agric. Handbook No. 552, U. S. Govt. Print Off., Washington, D.C.

Alcock, J. (1975). Social interactions in the solitary wasp *Cerceris simplex* (Hymenoptera: Sphecidae). *Behavior* 54, 142–152.

Alexander, R. D. (1974). The evolution of social behavior. *Annu. Rev. Ecol. Syst.* 325–379.

Allee, W. C. (1931). *Animal Aggregations: A Study in General Sociology.* Univ. Chicago Press, Chicago.

Allee, W. C., O. Park, A. E. Emerson, T. Park, and K. P. Schmidt. (1950). *Principles of Animal Ecology.* Saunders, Philadelphia.

Allee, W. C., and J. Wilder. (1939). Group protection for *Euplanaria dorotocephala* from ultraviolet radiation. *Physiol. Zool.* 12, 110–135.

Baroni-Urbani, C. (1979). Territoriality in social insects. In *Social Insects* (H. R. Herrman, ed.), pp. 91–120. Academic Press, New York.

Barth, R. H., L. J. Lester, P. Sroka, T. Kessler, and R. Hearn. (1975). Juvenile hormone promotes dominance behavior and ovarian development in social wasps (*Polistes annualris*). *Experientia* 31, 691–692.

Batra, S. W. T. (1978). Aggression, territoriality, mating and nest aggregation of some solitary bees (Halictidae, Megachilidae, Colletidae, Anthophoridae). *J. Kans. Entomol. Soc.* 51, 547–559.

Bernstein, I. S. (1976). Dominance, aggression and reproduction in primate societies. *J. Theor. Biol.* 60, 459–472.

Blancheteau, M. (1975). Observations sur le comportement aggressif de la Jourmi *Messor barbara* L. *Insectes Soc.* 22, 113–115.

Blum, M. S., and H. R. Hermann. (1978a). Venoms and venom apparatuses of the Formicidae: Myrmeciinae, Ponerinae, Dorylinae, Pseudomyrmecinae, Mrymicinae, and Formicinae. In *Arthropod Venoms* (S. Bettini, ed.), *Handbook of Experimental Pharmacology*, Vol. 48, pp. 871–894. Springer-Verlag, Berlin and New York.

Blum, M. S., and H. R. Hermann. (1978b). Venoms and venom apparatuses of the Formicidae: Dolichoderinae and Aneuretinae. In *Arthropod Venoms* (S. Bettini, ed.), *Handbook of Experimental Pharmacology*, Vol. 48, pp. 871–894. Springer-Verlag, Berlin and New York.

Blum, M. S., H. M. Fales, K. W. Tucker, and A. M. Collins. (1978). Chemistry of the sting apparatus of the worker honeybee. *J. Apic. Res.* 17, 218–221.

Bohm, M. K. (1972). Effects on environment and juvenile hormones on ovaries of the wasp, *Polistes netricus. J. Insect. Physiol.* 18, 1875–1883.

Bouillon, A. (1970). Termites of the Ethiopian region. In *Biology of Termites* (K. Krishna and F. M. Weesner, eds.), Vol. 2, pp. 153–280. Academic Press, New York.

Bristowe, W. S. (1932). Insects and other invertebrates for human consumption in Siam. *Trans. Entomol. Soc. London* 80, 387–404.

Brower, L. P., J. V. Z. Brower, and J. M. Corvino. (1967). Plant poisons in a terrestrial food chain. *Proc. Nat. Acad. Sci. USA*, 57, 893–898.

Brown, W. L., Jr. (1967). A new *Pheidole* with reversed phragmosis (Hymenoptera: Formicidae). *Psyche* 74, 331–339.

Brown, W. L., Jr. (1976). Contributions toward a reclassification of the Formicidae. Part VI. Ponerinae, tribe Ponerini, subtribe Odontomachiti. Section A. Introduction, subtribal characters. Genus *Odontomachus. Stud. Entomol.* 19, 67–171.

Brown, J. L., and G. H. Orians. (1970). Spacing patterns in mobile animals. *Annu. Rev. Ecol. Syst.* 1, 239–262.

Brown, W. L., Jr., and E. O. Wilson (1959). The evolution of dacetine ants. *Q. Rev. Biol.* 34, 278–294.

Bugnion, E. (1922). La guerre des fourmis et des termites, la genése des instincts expliquée par cette guerre. In *Le monde social des fourmis* (A. Forel, ed.), Vol. III, Appendix, pp. 173–225. Geneva.

Buren, W. F., H. R. Hermann, and M. S. Blum. (1970). The widespread occurrence of mandibular grooves in aculeate Hymenoptera. *J. Ga. Entomol. Soc.* 5, 185–196.

Buren, W. F., M. A. Naves, and T. C. Carlysle. (1977). False phragmosis and apparent specialization for subterranean warfare in *Pheidole lamia* Wheeler (Hymenoptera: Formicidae). *J. Ga. Entomol. Soc.* 12, 100–108.

Buskirk, R. E. (1981). Sociality in the Arachnida. In *Social Insects* (H. R. Hermann ed.), pp. 281–367. Academic Press, New York.

Butler, C. G., and J. B. Free. (1952). The behavior of worker honeybees at the hive entrance. *Behavior* 4, 263–292.

Chadab-Crepet, R., and Rettenmeyer, C. W. (1982). Comparative behavior of social wasps when attacked by army ants or other predators and parasites. In *The Biology of Social Insects* (M. D. Breed, C. D. Michener, and H. E. Evans, eds.) Proc. 9th Inter. Cong. Stud. Soc. Insects, Boulder, Colorado.

Corn, M. L. (1972). Notes on the biology of *Polistes carnifex* (Hymenoptera: Vespidae) in Costa Rica and Colombia. *Psyche* 79, 150–157.

Cott, H. B. (1940). *Adaptive Coloration in Animals.* Oxford Univ. Press, London and New York.

Craig, R. (1979). Parental manipulation, kin selection, and the evolution of altruism. *Evolution* 33, 319–334.

Crozier, R. H. (1979). Genetics of sociality. In *Social Insects* (H. R. Hermann, ed.), pp. 223–286. Academic Press, New York.

Daly, H. V. (1953). A comparative survey of the sting of aculeate Hymenoptera. Thesis, Univ. of Kansas, Lawrence.

Darwin, C. R. (1859). *On the Origin of Species by Means of Natural Selection, or the Preservation of Favored Races in the Struggle for Life*, 1st ed. Murray, London.

Decelles, P. (1976). Studies on the behavioral ecology of *Pheidole bicarinata vinelandica* Forel. (Hymenoptera: Formicidae). M. S. Thesis, Univ. of Georgia, Athens.

Deleurance, E.-P. (1957). Contribution à l'étude biologique des Polistes (Hymenopteres Vespoides): I, l'activité de construction. *Ann. Sci. Nat. Zool.* (11th ser.) 19, 91–222.

Delfino, G., M. T. Marino Piccioli, and C. Callino. (1979). Fine structure of the glands of Van der Vecht's organ in *Polistes gallicus* (L.) (Hymenoptera Vespidae). *Monit. Zool. Ital.* 13, 221–247.

Deligne, J., A. Quennedey, and M. S. Blum. (1981). The enemies and defense mechanisms of termites. In *Social Insects* (H. R. Hermann, ed.), pp. 1–76. Academic Press, New York.

Dodson, C. H. (1966). Ethology of some bees of the tribe Euglossini (Hymenoptera: Apidae). *J. Kans. Entomol. Soc.* 39, 607–629.

Drory, E. (1872). Einige beobachtungen an *Melipona scutellaris*. *Bienen-Ztg. Schweiz* 28, 157–206.

Eberhard, W. G. (1972). Altruistic behavior in a sphecid wasp: Support for kin-selection theory. *Science* 172, 1390–1391.

Edery, H., J. Ishary, S. Gitter, and H. Joshua. (1978). Venoms of Vespidae. In *Arthropod Venoms* (S. Bettini, ed.). *Handbook of Experimental Pharmacology*, Vol. 48, pp. 691–771. Springer-Verlag, Berlin and New York.

Edmunds, M. (1974). "Defense in Animals." Longman, Harlow, England.

Edwards, R. (1980). *Social Wasps, Their Biology and Control*. Rentokil Ltd., East Grinstead, England.

Eickwort, G. C. (1981). Presocial insects. In *Social Insects* (H. R. Hermann, ed.) pp. 199–280. Academic Press, New York.

Eickwort, G. C., K. R. Eickwort, and E. G. Linsley. (1977). Observations on nest aggregations of the bees *Diadasia olivacea* and *D. diminuta* (Hymenoptera: Anthophoricidae). *J. Kans. Entomol. Soc.* 50, 1–17.

Emerson, A. E. (1959). Social insects. In *Encyclopaedia Britannica*, Vol. 20, pp. 871–878. Benton, Chicago.

Emlen, S. T. (1981). Altruism, kinship, and reciprocity in the white-fronted bee-eater. In *Natural Selection and Social Behavior* (R. D. Alexander and D. W. Tinkle, eds.), pp. 217–230. Chiron Press, New York.

Evans, H. E. (1958). The evolution of social life in wasps. *20th Proc. Int. Congr. Entomol.* (Montreal, 1956), 2, 449–457.

Evans, H. E., and M. J. West Eberhard. (1970). *The Wasps*. Univ. Michigan Press, Ann Arbor.

Fellers, J. H., and G. M. Fellers. (1976). Tool use in an ant and its implications for competitive interactions. *Science* 192, 70–71.

Fletcher, D. J. C. (1978). The African bee, *Apis mellifera adansonii*, in Africa. *Annu. Rev. Entomol.* 23, 151–171.

Free, J. B. (1970). The behavior of wasps (*Vespula germanica* L. and *V. vulgaris* L.) when foraging. *Insectes Soc.* 17, 11–20.

Free, J. B., and C. G. Butler. (1959). *Bumblebees*. New Naturalist, Collins, London.

Friese, H. (1914). Die Dienenfauna von Java. *Tjdschr. Entomol.* 57, 43–58.

Frings, H., and M. Frings. (1975). *Animal Communication*. Univ. Oklahoma Press, Norman.

Gary, N. E. (1974). Pheromones that affect the behavior and physiology of honeybees. In *Pheromones* (M. C. Birch, ed.), pp. 200–221. North-Holland Publ., Amsterdam.

Gaul, A. T. (1953). Additions to vespine biology. XI. Defence flight. *Bull. Brooklyn Entomol. Soc.* 48, 35–37.

Gentry, J. B. (1974). Response to predation by colonies of the Florida harvester ant, *Pogonomyrmex badius. Ecology* 55, 1328–1338.

Gerstell, R. (1939). Certain mechanics in winter quail losses revealed by laboratory experimentation. *Trans. North Am. Wildl. Inst.* (4th), 462–467.

Gibo, D. L. (1974). A laboratory study on the selective advantage of foundress associations in *Polistes Fuscatus. Can. Entomol.* 106, 101–106.

Gibo, D. L. (1978). The selective advantage of foundress associations in *Polistes fuscatus* (Hymenoptera: Vespidae): A field study of the effects of predation on productivity. *Can. Ent.* 110, 519–540.

Gotwald, W. H. (1974). Predatory behavior and food preferences of driver ants in selected African habitats. *Ann. Entomol. Soc. Am.* 67, 877–886.

Gotwald, W. H. (1976). Behavioral observations on army ants of the genus *Aenictus* (Hymenoptera: Formicidae). *Biotropica* 8, 59–65.

Grasse, P.-P., and C. Noirot. (1948). Sur le nid et la biologie de *Sphaerotermes sphaerothorax* (Sjost). Termite constructurer de meules sans champignons. *Ann. Sci. Nat. Zool. Biol. Anim.* (11) 10, 149–166.

Green, A., R. D. Akre, and P. Landolt. (1976). The aerial yellowjacket, *Dolichovespula arenaria* (Fab.): Nesting biology, reproductive production, and behavior (Hymenoptera: Vespidae). *Melanderia* 26, 1–34.

Guiglia, D. (1972). Les guêpes sociales (Hymenoptera Vespidae) d'Europe occidentale et septentrionale. Masson et Cie, Paris.

Hamilton, W. D. (1963). The evolution of altruistic behavior. *Am. Natur.* 97, 354–356.

Hamilton, W. D. (1964). The genetical evolution of social behavior. I and II. *J. Theor. Biol.* 7, 1–52.

Hamilton, W. D. (1972). Altruism and related phenomena, mainly in social insects. *Annu. Rev. Ecol. Syst.* 3, 193–232.

Hegh, E. (1922). Les termites. Imprimeries Industrielle et Financiére Bruxelles, Belgium.

Henry, C. S. (1972). Eggs and repagula of *Ululodes* and *Ascaloptynx* (Neuroptera: Ascalaphidae): A comparative study. *Psyche* 79, 1–22.

Hermann, H. R. (1968). The hymenopterous poison apparatus. VII. *Simopelta oculata* (Hymenoptera: Formididae: Ponerinae). *J. Ga. Entomol. Soc.* 3, 163–166.

Hermann, H. R. (1971). Sting autotomy, a defensive mechanism in certain social Hymenoptera. *Insectes Soc.* 18, 111–120.

Hermann, H. R. (1975). The ant-like venom apparatus of *Typhoctes peculiaris*, a primitive mutillid wasp. *Ann. Entomol. Soc. Am.* 68, 882–884.

Hermann, H. R. (1979). Insect sociality—An introduction. In *Social Insects* (H. R. Hermann, ed.), pp. 1–33. Academic Press, New York.

Hermann, H. R., R. Baer, and M. Barlin. (1975). Histology and function of the venom gland system in formicine ants. *Psyche* 82, 67–73.

Hermann, H. R., and M. S. Blum. (1968). The hymenopterous poison apparatus. VI. *Camponotus pennsylvanicus* (Hymenoptera: Formicidae). *Psyche* 75, 216–227.

Hermann, H. R., and M. S. Blum. (1981). Defensive mechanisms in the social Hymenoptera. In *Social Insects* (H. R. Hermann, ed.), pp. 77–197. Academic Press, New York.

Hermann, H. R., and T. F. Dirks. (1974). Sternal glands in polistine wasps: Morphology and associated behavior. *J. Ga. Entomol. Soc.* 9, 1–8.

Hermann, H. R., and T. F. Dirks. (1975). Biology of *Polistes annularis* (Hymenoptera: Vespidae). I. Spring Behavior. *Psyche* 82, 97–108.

Hermann, H. R., A. N. Hunt, and W. F. Buren. (1971). Mandibular gland and mandibular groove in *Polistes annularis* (L.) and *Vespula maculata* (L.) (Hymenoptera: Vespidae). *Int. J. Insect Morphol. Embryol.* 1, 43–49.

Hermann, H. R., and A. Young. (1980). Artificially elicited defensive behavior and reciprocal intraspecific aggression in *Paraponera clavata* (Hymenoptera: Formicidae: Ponerinae). *J. Ga. Entomol. Soc.* 15, 8–10.

Hinde, R. A. (1970). *Animal Behavior, a Synthesis of Ethology and Comparative Psychology.* McGraw-Hill, New York.

Hockings, H. (1884). Notes on two Australian species of *Trigona*. *Trans. Entomol. Soc. London* 32, 149–150.

Hölldobler, B. (1974). Home range orientation and territoriality in harvesting ants. *Proc. Natl. Acad. Sci. USA* 71, 3274–3277.

Hölldobler, B. (1976). The behavioral ecology of mating in harvester ants (Hymentoptera: Formicidae) *Pogonomyrmex*. *Behav. Ecol. Sociobiol.* 1, 405–423.

Hölldobler, B. (1977). Communication in social Hymenoptera. In *How Animals Communicate* (T. A. Sebeok, ed.), pp. 423–471. Indiana Univ. Press, Bloomington.

Hölldobler, B. (1978). Ethological aspects of chemical communication in ants. *Adv. Study Behav.* 8, 75–115.

Hölldobler, B. (1979). Territoriality in ants. *Proc. Am. Philos. Soc.* 123, 211–218.

Hölldobler, B., and E. O. Wilson. (1970). Recruitment trails in the harvester ant, *Pogonomyrmex badius*. *Psyche* 77, 385–399.

Hölldobler, B., and E. O. Wilson. (1977a). Weaver ants: Social establishment and maintenance of territory. *Science* 195, 900–902.

Hölldobler, B., and E. O. Wilson. (1977b). Colony-specific territorial pheromone in the African weaver ant *Oecophylla longinoda* (Latreille). *Proc. Natl. Acad. Sci. USA* 74, 2072–2075.

Hung, A. C. F., and W. L. Brown. (1964). Structure of gastric apex as a subfamily character of the Formicinae (Hymenoptera: Formicidae). *J. N. Y. Entomol. Soc.* 74, 198–200.

Ishay, J. (1973). Thermoregulation by social wasps: Behavior and phero-mones. *Trans. N. Y. Acad. Sci.* (2) 36, 447–462.

Ishay, J., H. Bytinski-Salz, and A. Shulov. (1967). Contributions to the bionomics of the Oriental hornet (*Vespa orientalis* Fab.). *Isr. J. Entomol.* 2, 45–106.

Ishay, J., and E. M. Landau (1972). Vespa larvae send out rhythmic hunger signals. *Nature* 237, 286–287.

Iwata, K. (1967). Report on the fundamental research on the biological control of insect pests in Thailand: II, the report on the bionomics of subsocial wasps of Stenogastrinae (Hymenoptera, Vespidae). *Nat. Life Southeast Asia* 5, 259–293.

Iwata, K. (1969). On the nidification of *Ropalidia* (Anthreneida) *taiwana kochunensis* Sonan in Formosa (Hymenoptera, Vespidae). *Kontyu* 37, 367–372.

Iwata, K. (1971). *Evolution of Instinct*. Amerind Publ. Co., New Delhi.

Janzen, D. H. (1967). Interaction of the bull's-horn acacia (*Acacia cornigera* L.) with an ant inhabitant (*Pseudomyrmex ferruginea* F. Smith) in eastern Mexico. *Univ. Kansas Sci. Bull.* 47, 315–358.

Jeanne, R. L. (1970). Chemical defense of brood by a social wasp. *Science* 168, 1465–1466.

Jeanne, R. L. (1972). Social biology of the neotropical wasp *Mischocyttarus drewseni. Bull. Mus. Comp. Zool.* 144, 63–150.

Jeanne, R. L. (1975). The adaptiveness of social wasp architecture. *Q. Rev. Biol.* 50, 267–287.

Jeanne, R. L. (1977). Behavior of the obligate social parasite *Vespula arctica* (Hymenoptera: Vespidae). *J. Kans. Entomol. Soc.* 50, 541–557.

Jeanne, R. L. (1978). Intraspecific nesting associations in the Neotropical social wasp *Polybia rejecta* (Hymenoptera: Vespidae). *Biotropica* 10, 234–235.

Jeanne, R. L. (1979a). Construction and utilization of multiple combs in *Polistes canadensis* in relation to the biology as a predaceous moth. *Behav. Ecol. Sociobiol.* 4, 293–310.

Jeanne, R. L. (1979b). Nest of the wasp *Clypearia neyranchi* (Hymenoptera, Vespidae). *J. N. Y. Entomol. Soc.* 87, 78–84.

Jeanne, R. L. (1980). Evolution of social behavior in the Vespidae. *Annu. Rev. Entomol.* 25, 371–396.

Jeanne, R. L. (1981). Chemical communication during swarm emigration in the social wasp *Polybia sericea* (Olivier). *Anim. Behav.* 29, 102–113.

Kemper, H., and E. Dohring. (1967). Die sozialen Faltenwespen Mittel-ewropas. Paul Parey, Berlin.

Kerr, W. E., and C. da C. Cruz. (1961). Funcoes diferentes tomadas pela glandula mandibular na evolucae das abelhas em geral e em "Trigona (Oxytrigona) tataira" em especial. *Rev. Bras. Biol.* 21, 1–16.

Kerr, W. E., and E. de Lello. (1962). Sting glands in stingless bees—A vestigial character (Hymenoptera: Apidae). *J. N. Y. Entomol. Soc.* 70, 190–214.

Konishi, M. (1971). Ethology and neurobiology. *Am. Sci.* 59, 56–63.

Krishna, K., and F. M. Weesner. (1969). *Biology of Termites*, Vol. I. Academic Press, New York.

Krishna, K. and F. M. Weesner. (1970). *Biology of Termites*, Vol. II. Academic Press, New York.

Landolt, P. J., and R. D. Akre. (1979). Occurrence and location of exocrine glands in some social Vespidae (Hymenoptera). *Ann. Entomol. Soc. Am.* 72, 141–148.

Landolt, P. J., R. D. Akre, and A. Green. (1977). Effects on colony division on *Vespula atropilosa* (Sladen) (Hymenoptera: Vespidae). *J. Kans. Entomol. Soc.* 50, 135–147.

Lecomte, M. J. (1951). Recherches sur le comportment agressif des ouvrieres *d'Apis mellifica. Behaviour* 4, 60–66.

Lin, N., and C. D. Michener. (1972). Evolution of sociality in insects. *Q. Rev. Biol.* 47, 131–159.

Lindauer, M. (1967). Recent advances in bee communication and orientation. *Annu. Rev. Entomol.* 12, 439–470.

Little, M. (1979). *Mischocyttarus flavitarsis* in Arizona: Social and nesting biology of a polistine wasp. *Zts. Tirepslychol.* 50, 282–312.

Lorenz, K. (1974). *On Aggression.* Harcourt Brace Jovanovich, New York.

Maidl, F. (1934). *Diedefensge wohnheiten und Institute der Staattenbildenden Insekton*, pp. 316–317. Vergl. Wagner, Vienna.

Maier, R. A., and B. M. Maier. (1970). *Comparative Animal Behavior.* Brooks Cole, Belmont, Calif.

Mariano, J. (1910). Sobre os meios naturaes de defesa das abelhas sem ferrao. *Chacaras Quintaes, Sao Paulo* 1, 50–53.

Marino Piccioli, M. T., and L. Pardi. (1970). Studi sulla biologia de *Belenogaster griseus* (Fab.). *Monit. Zool. Ital., NS (Suppl.* III) 9, 197–225.

Markl, H. (1965). Stridulation in leaf-cutting ants. *Science* 149, 1392–1393.

Markl, H. (1967). Die Verstandigung durch Stridulationssignale bei Blatt-schneiderameisen. I. Die biologische Bedeutung der Stridulation. *Z. Vergl. Physiol.* 57, 299–330.

Markl, H. (1968). Die Verstandigung durch Stridulationssignale bei Blatt-schneiderameisen. II. Erzengung und Eigenschaften der Signale. *Z. Vergl. Physiol.* 60, 103–150.

Markl, H. (1973). The evolution of stridulatory communication in ants. *7th Proc. Int. Cong. Int. Union Study Soc. Insects*, 1973. pp. 258–265.

Markl, H., and S. Fuchs. (1972). Kopfsignale mit Alarmfunktion bei Rossa-meisen (*Camponotus*) (Formicidae, Hymenoptera). *Z. Vergl. Physiol.* 76, 204–225.

Maschwitz, U. (1964). Gefahrenalarmstoffe und Gefahrenalarmierung bei sonzialen Hymenopteron. *Z. Vergl. Physiol.* 47, 596–655.

Maschwitz, U., and W. Kloft. (1971). Morphology and function of the venom apparatus of insects, bees, wasps, ants and caterpillars. In *Venomous Animals and Their Venoms* (W. Bucherl and E. E. Buckley, eds.), Vol. 3, Chapter 44, pp. 31–60. Academic Press, New York.

Maschwitz, U., and E. Maschwitz. (1974). Platyende Arbeiterinnen: Eine neue Art der Feindabwehr bei sozialen Hautfluglern. *Oecologia* 14, 289–294.

Matthews, R. W. (1968). *Microstigmus comes*: Sociality in a sphecid wasp. *Science* 160, 787–788.

Matthews, R. W. (1970). A new thrips-hunting *Microstigmus* from Costa Rica (Hymenoptera: Sphecidae: Pemphredoninae). *Psyche* 77, 120–126.

Mayr, G. (1892). *Derpanognathus rugosus* Mayr. *Termeszettud. Fuz.* 15, 127.

Mayr, G. (1893). Eryanzende Bemerkungen zu E. Wasmann's Artikel uber Springende Ameisen. *Wien. Entomol. Ztg.* 12, 23.

Michener, C. D. (1962). Biological observations on the primitively social bees of the genus *Allodapula* in the Australian region (Hymenoptera, Xylocopinae). *Insectes Soc.* 9, 355–373.

Michener, C. D., and D. J. Brothers. (1974). Were workers of eusocial Hymenoptera initially altruistic or oppressed? *Proc. Natl. Acad. Sci. USA* 71, 671–674.

Michener, C. D., R. B. Lange, J. J. Bigarella, and R. Salamuni. (1958). Factors influencing the distribution of bees' nests in earth banks. *Ecology* 39, 207–217.

Miller, R. C., and F. E. Kurczewski. (1973). Intraspecific interactions in aggregations of *Lindenius* (Hymenoptera: Sphecidae, Crabroninae). *Insectes Soc.* 20, 365–378.

Moore, B. P. (1969). Biochemical studies in termites. In *Biology of Termites* (K. Krishna and F. M. Weesner, eds.), Vol. 1, pp. 407–432. Academic Press, New York.

Mosebach-Pukowski, E. (1937). Über die Raupengesellschaften von *Vanessa io* und *Vanessa urticae*. *Z. Morph. Okol. Tiere* 33, 358–380.

Myers, J., and M. D. Loveless. (1976). Nesting aggregations of the euglossine bee *Euplusia surinamensis* (Hymenoptera: Apidae): Individual interactions and the advantage of living together. *Can. Entomol.* 108, 1–6.

Noirot, C. (1970). The nests of termites. In *Biology of Termites* (K. Krishna and F. M. Weesner, eds.), Vol. 2, pp. 73–125. Academic Press, New York.

O'Connor, R., W. Rosenbrook, and R. Erickson. (1963). Hymenoptera: Pure venom from bees, wasps and hornets. *Science* 139, 420.

Oster, G. F., and E. O. Wilson. (1978). Caste and ecology in the social insects. Princeton Univ. Press, Princeton, N. J.

Owen, J. (1962). The behavior of the social wasp *Polistes fuscatus*. (Vespidae) at the nest, with special reference to differences between individuals. Ph.D. thesis, Univ. Michigan.

Pardi, L. (1948). Dominance order in *Polistes* wasps. *Physiol. Zool.* 21, 1–13.

Pardi, L., and M. T. Marino Piccioli. (1970). Studi sulla biologia di *Belonogaster*. 2. Differenziamento castale incipiente in *B. griseus* (Fab.). *Monit. Zool. Ital.* (NS) (*Suppl.* 3), 235–265.

Piccioli, M. T. M., and L. Pardi. (1970). Studi della biologia di *Belonogaster* (Hymenoptera, Vespidae): I. sull'etogramma di *Belonogaster griseus* (Fab.). *Mon. Zool. Ital.*

Pliner, P., L. Krames, and T. Alloway. (1975). "Advances in the Study of Communication and Effect," Vol. 2. Plenum, New York.

Post, D. C., and R. L. Jeanne. (1980). Morphology of the sternal glands of *Polistes fuscatus* and *P. canadensis* (Hymenoptera, Vespidae). *Psyche* 87, 49–58.

Post, D. C., and R. L. Jeanne. (1981). Colony defense against ants by *Polistes fuscatus* (Hymenoptera: Vespidae) in Wisconsin. *J. Kans. Entomol. Soc.* 54, 599–615.

Potter, N. B. (1965). Some aspects of the biology of *Vespula vulgaris* L. Ph.D. Thesis, Univ. Bristol, England.

Powell, G. C., and R. B. Nickerson. (1965). Aggregations among juvenile king crabs (*Paralithodes comtschatica*, Tilesius), *Anim. Behav.* 13, 374–380.

Pursley, R. E. (1973). Stinging Hymenoptera. *Am. Bee J.* 113, 131–132.

Rau, P. (1931). The nesting habits of *Polistes rubiginosis* with special reference to pleometrosis in this and other species of *Polistes* wasps. *Psyche* 38, 129–144.

Rau, P., and N. Rau (1970). *Wasp Studies Afield.* Dover, New York.

Reitschel, P. (1937). Bau und Funktion des Wehrstachels der staatenbildenden Bienen und Wespen. *Z. Morphol. Oekol. Tiere* 33, 313–357.

Rettenmeyer, C. W. (1963). Behavioral studies of army ants. *Univ. Kansas Sci. Bull.* 44, 281–465.

Ribbands, C. R. (1964). *The Behavior and Social Life of Honey Bees.* Dover, New York.

Richards, O. W. (1953). *The Social Insects.* MacDonald, London (reprinted by Harper, New York).

Richards, O. W. (1965). Concluding remarks on the social organization of insect communities. *Symp. Zool. Soc. London* 14, 169–172.

Richards, O. W., and M. J. Richards. (1951). Observations on the social wasps of South America (Hymenoptera: Vespidae). *Trans. R. Entomol. Soc. London* 14, 169–172.

Roberts, R. B., and C. H. Dodson. (1967). Nesting biology of two communal bees, *Euglossa imperialis* and *Euglossa ignita* (Hymenoptera: Apidae), including description of larvae. *Ann. Entomol. Soc. Amer.* 60, 1007–1014.

Rodriguez, V. M. (1968). Estudo sobre Vespas sociaes do Brasil (Hymenoptera-Vespidae). Doctoral Thesis Fac. Filos. Cienc. Letras de Rio Claro, Universidad de Campinas.

Roseler, P. F., I. Roseler, and A. Strambi. (1980). The activity of corpora allata in dominant and subordinated females of the wasp *Polistes gallicus. Insectes Soc.* 27, 97–107.

Rubenstein, D. I., and R. W. Wrangham. (1980). Why is altruism towards kin so rare? *Z. Tierpsychol.* 54, 381–387.

Sakagami, S. F. (1960). Preliminary report on the specific difference of behavior and other ecological characters between European and Japanese honey bees. *Acta Hymenopt.* 1, 171–198.

Sakagami, S. F. (1976). Specific differences in the bionomic characters of bumblebees. A comparative review. *J. Fac. Sci. Hokkaido Univ.* 20, 390–447.

Sakagami, S. F., and Y. Akahira. (1960). Studies on the Japanese honeybee, *Apis cerana cerana* Fabricus: 8. Two opposing adaptations in the poststinging behavior of honeybees. *Evolution* 14, 29–40.

Sakagami, S. F., and C. D. Michener. (1962). *The Nest Architecture of the Sweat Bees (Halictinae): A Comparative Study of Behavior.* Univ. Kansas Press, Lawrence.

Sakagami, S. F., and R. Zucchi. (1965). Winterverhalten einer neotropischen Hummel, *Bombus atratus*, innerhalb des Beobachtungskastens: ein Bertrag zur Biologie der Hummeln. *J. Fac. Sci.* (Hokkaido Univ., 6th ser.) (Zool.) 15, 712–762.

Saslavasky, H. J. Ishay, and R. Ikan. (1973). Alarm substances as toxicants in the Oriental hornet colony. *Life Sci.* 12, 135–144.

Schmidt, J. O. (1981). Biochemistry of insect venoms. *Annu. Rev. Entomol.* Vol. 27, 339–368.

Schmitt, C. (1920). Beitrage zur Biologie der Feldwespa (*Polistes gallicus* L.). *Zts. Wiss. Insektenbiol.* 16, 146–161.

Schremmer, F. (1972). Beobachtungen zur Biologie von *Apoica pallida* (Olivier, 1971), einer neotropischen sozialen Faltenwespe (Hymenoptera, Vespidae). *Insectes Soc.* 19, 343–357.

Shearer, D. A., and R. Boch. (1965). 2-Heptanone in the mandibular gland secretion of the honey bee. *Nature (London)* 206, 530.

Sladen, F. W. L. (1912). *The Humble-bee, its Life-history and how to Domesticate it, with Descriptions of all the British Species of Bombus and Psithyrus.* Macmillan Co., London.

Smith, J. M. (1978). The evolution of behavior. In *Evolution*, pp. 92–101. W. H. Freeman, San Francisco.

Snelling, R. R. (1981). Systematics of social Hymenoptera. In *Social Insects* (H. R. Hermann, ed.), Vol. II, pp. 369–453. Academic Press, New York.

Spradbery, J. P. (1973). *Wasps.* Univ. Wash. Press, Seattle.

Spangler, H. G. (1967). Ant stridulations and their synchronization with abdominal movement. *Science* 155, 1687–1689.

Spieth, H. T. (1948). Notes on a colony of *Polistes fuscatus hunteri*, Bequaert. *J. N. Y. Entomol. Soc.* 56, 155–169.

Starr, C. K. (1979). Origin and evolution of insect sociality: A review of modern theory. In *Social Insects* (H. R. Hermann, ed.). pp. 35–89. Academic Press, New York.

Starr, C. K. (1981). Defensive tactics of social wasps. Ph.D. Thesis. Univ. Georgia.

Steiner, A. (1932). Die Arbeitsteilung der Feldwespe *Polistes dubia* Kohl. *Zts. vgl. Physiol.* 17, 101–152.

Steiner, A. L. 1975. Description of the territorial behavior of *Podalonia valida* (Hymenoptera, Sphecidae) females in Southeast Arizona, with remarks on digger wasp territorial behavior. *Quaest. Entomol.* 11, 113–127.

Stephen, W. P., G. E. Bohart, and P. F. Torchio. (1969). *The Biology and External Morphology of Bees.* Oregon State Univ., Agric. Exp. Stn., Corvallis.

Stevcic, Z. (1971). Laboratory observations on the aggregations of the spiny spider crab (*Maja squinado* Herbst). *Anim. Behav.* 19, 18–25.

Strassmann, J. E. (1981). Wasp reproduction and kin selection: Reproductive competition and dominance hierarchies among *Polistes annularis* foundresses. *Florida Entomol.* 64, 74–88.

Stuart, A. M. (1969). Social behavior and communication. In *Biology of Termites* (K. Krishna and F. M. Weesner, eds.), pp. 193–232. Academic Press, New York.

Taylor, R. W. (1978). *Nothomyrmecia macrops*: A living fossil-ant rediscovered. *Science* 201, 979–985.

Thomas, C. R. (1960). The European Wasp (*Vespula germanica* F.) in New Zealand. *Inf. Ser. Dep. Sci. Ind. Res. N. Z.* 27, 1–73.

Tinbergen, N. (1974). *The Study of Instinct*. Oxford University Press, New York.

Turillazzi, S. (1979). Tegumental glands in the abdomen of some European *Polistes* (Hymenoptera: Vespidae). *Monit. Zool. Ital.* (NS) 12, 72.

Turillazzi, S., and A. Ugolini. (1978). Nest defense in European *Polistes* (Hymenoptera: Vespidae). *Monit. Zool. Ital.* [N.S.] 12, 72.

Weber, N. A. (1972). "Gardening Ants: the Attines." *Mem. Amer. Phil. Soc.* (no. 92) Amer. Phil. Soc., Philadelphia, 146 pp.

West Eberhard, M. J. (1969). The social biology of polistine wasps. *Univ. Mich. Mus. Zool. Misc. Publ.* 140, 1–101.

West Eberhard, M. J. (1979). Sexual selection, social competition, and evolution. *Proc. Amer. Phil. Soc.* I, II, III, 222–234.

West Eberhard, M. J. (1981). Intragroup selection and the evolution of insect societies. In *Natural Selection and Social Behavior* (R. D. Alexander and D. W. Tinkle, eds.), pp. 3–17. Blackwell Scientific Publications, London.

Weyrauch, W. K. (1928). Beitrag zur Biologie von *Polistes. Biol. Zbl.* 48, 407–427.

Weyrauch, W. K. (1935). *Dolichovespula* und *Vespa*. Vergleichende Ubersicht uber zwei wesentliche Lebenstvpn bie sozialen Wespen. Mit Bezugnahme auf die Frage nach der Fortschrittlichkeit tierischer Organisation. I. *Biol. Zbl.* 55, 484–524.

Wheeler, W. M. (1910). *Ants, Their Structure, Development and Behavior*. Columbia Univ. Press, New York.

Wheeler, W. M. (1928). *The Social Insects: Their Origin and Evolution*. Kegan Paul, Trench, Trubner, London.

Wheeler, W. M. (1936). Ecological relations of ponerine and other ants to termites. *Proc. Amer. Acad. Arts Sci.* 71, 159–243.

Williams, F. X. (1919). Philippine wasp studies: II, descriptions of new species and life history studies. *Bull. Exp. Sta. Hawaiian Sugar Planters Assn. Entomol. Ser.* 14, 19–184.

Wilson, E. O. (1971). *The Insect Societies*. Harvard Univ. Press, Cambridge, Mass.

Wilson, E. O. (1975). *Sociobiology, the New Synthesis*. Harvard Univ. Press, Cambridge, Mass.

Wood, T. G., and W. A. Sands. (1978). The role of termites in ecosystems. In *Production Ecology of Ants and Termites* (M. V. Brian, ed.), *International Biology Programme*, Vol. 13, pp. 245–292. Cambridge Univ. Press, London and New York.

Wroughton, R. C. (1892). Our ants. *J. Bombay Nat. Hist. Soc.* 7, 13–60, 175–202.

Wynn-Edwards, V. C. (1962). "Animal Despersion in Relation to Social Behavior," Oliver and Boyd, Edinburgh.

Yamane, S. (1971). Daily activities of the founding queens of two *Polistes* species, *P. snelleni* and *P. biglumis* in the solitary stage (Hymenoptera, Vespidae). *Kontyu* 39, 203–217.

Yamane, S. (1976). Morphological and taxonomic studies on vespine larvae, with reference to the phylogeny of the subfamily Vespinae (Hymenoptera: Vespidae). I. Studies on the vespoid larvae. *Insecta Matsumurana* (NS) 8, 1–45.

Yamane, S., and S. Makino. (1977). Bionomics of *Vespa analis insularis* and *V. mandarinia latilineata* in Hokkaido, Northern Japan, with notes on vespine embryo nests (Hymenoptera: Vespidae). *Insecta Matsumurana* (NS) 12, 1–33.

Yoshikawa, K. (1963). Introductory studies on the life economy of polistine wasps. II. Superindividual stage. 3. Dominance order and territory. *Jap. J. Ecol.* 14, 55–61.

Yoshikawa, K. (1964). Predatory hunting wasps as the natural enemies of insect pests in Thailand. *Nat. Life Southeast Asia (Tokyo)* 3, 391–398.

Yoshikawa, K., R. Ohgushi, and S. F. Sakagami. (1969). Preliminary report on entomology of the Osaka City University 5th Scientific Expedition to Southeast Asia, 1966, with descriptions of two new genera of stenogastrine wasps by J. van der Vecht. *Nat. Life Southeast Asia (Tokyo)* 6, 153–182.

Zucchi, R., S. F. Sakagami, and J. M. F. de Camargo. (1969). Biological observations on a Neotropical parasocial bee, *Eulaema nigrita*, with a review of the biology of Euglossinae: A comparative study. *J. Fac. Sci. Hokkaido Univ 6th ser. (Zool.)* 17, 271–380.

2
Defensive Mechanisms in
Polistes Wasps
Stefano Turillazzi

INTRODUCTION

Paper wasps of the genus *Polistes* are among the commonest social wasps. They are present in most parts of the globe (Yoshikawa, 1962). This distribution has enhanced the study of their social organization, though many uncertainties with regard to their adaptation to various environments remain to be solved. *Polistes* wasps have acquired efficient defensive mechanisms against nest predators and parasites that, together with the reduced seasonal cycle and cold in the temperate zones, are the principal selective forces.

Within colonies of social wasps are concentrations of very nourishing material. In the case of *Polistes*, the nests consist of a simple paper wrapping in which the defenseless future generations are neatly packaged, together with occasional sugary substances. Their stelo-cyttarous nests (attached to the substratum by a stem) lack both the protective envelope found in other Vespidae and the shelter afforded by hollows in the ground. Thus, their defense must depend on other means.

PREDATION

Predators Attacking *Polistes*

Richards and Richards (1951) and Jeanne (1975) indicate that ants and birds are the most important predators of social wasps and therefore of *Polistes*. In the tropics, ants destroy more colonies than do birds, while the opposite seems to occur in temperate zones. Other social wasps (with Vespinae in the lead) may destroy *Polistes* colonies, particularly in the oriental regions, while other occasional predators have also been reported. Table 2.1 updates Jeanne's (1975) list of predators and includes unpublished data for *Polistes*.

While the impact of predation on colonies of tropical *Polistes* has not yet been studied in depth, observations are available on *P. snelleni* (Yamane, 1969) and *P. biglumis* (Yamane and Kawamichi, 1975) from northern Japan, *P. fuscatus* (Gibo, 1978) and *P. apachus* (Gibo and Metcalf, 1978) from North America, and *P. nimpha* (Turillazzi and Cervo, in prep.) from central Italy. These studies were made in places where parasitism is scarce. The survival rate of *P. nimpha* colonies (haplometrotic* nest-founding species) during their active season is shown in Figure 2.1. Similarly, *P. snelleni P. biglumis* (both haplometrotic species) and *P. fuscatus* (haplometrotic and pleometrotic†) all suffer their greatest losses prior to worker emergence, owing to predation on foundresses as well as brood, particularly when the first larvae begin to pupate. An exception to this is *P. apachus*, in which maximum losses occur somewhat earlier. Colony destruction is due mostly to birds and, in Japan, to foxes. Losses can be very heavy: reproductives were produced in only 1 out of 37 colonies of *P. nimpha*, 3 out of 38 colonies of *P. snelleni* (Yamane, 1969), and 11 out of 40 colonies of *P. biglumis* (Yamane and Kawamichi, 1975).

Defensive Mechanisms Against Predators

Social wasps use several defenses against such highly organized invertebrate predators as ants (Richards and Richards, 1951; Jeanne, 1975): (1) construction of their nests in places where predaceous ants are absent or infrequent; (2) reduction in the total length of colony cycle and thus in the period of worker production; (3) duration of the reproductive capacity so that, at any given moment, only a small quantity of brood is exposed to attacks; and (4) protection by direct defenses. The same strategies can be used against vertebrate predators as well. In this case mimetic nidification should be added to the list. This strategy, useless against ants, is most effective against animals that locate prey by sight, such as birds and mammals.

Polistes wasps utilize all these defensive strategies.

---

*Nest found by a single foundress female.

†Nest found by two or more associate foundresses.

TABLE 2.1. Predators of *Polistes* Wasps

Predators	Prey	Location	Source
Ants			
Camponotus fallax	*P. pallipes*	Missouri	Rau (1930a)
Dolichoderinae ants	*P. spilophorus*	Ghana	Richards (1969)
Eciton vagans	*Polistes* sp.	Paraguay	Fiebrig (1907)
Monomorium ebeninum	*P. cinerascens*	Costa Rica	Windsor (in Jeanne, 1975)
Pheidole sp.	*P. cinerascens, P. canadensis*	Costa Rica	Windsor (in Jeanne, 1975)
Crematogaster scutellaris	*P. gallicus, P. foederatus*	Italy	Turillazzi and Ugolini (1979)
	P. nimpha		
Pheidole sp.	*P. nimpha*	Italy	Turillazzi and Ugolini (1979)
Lasius sp.	*P. nimpha*	Italy	Turillazzi and Ugolini (1979)
Plagiolepis sp.	*P. nimpha*	Italy	Turillazzi and Ugolini (1979)
"Ants"	*P. chinensis antennalis*	Japan	Suzuki (1978)
"Ants" various species	*P. fuscatus hunteri*	Florida	Spieth (1948)
Pheidole dentata	*Polistes* sp.	Texas	Reed and Vinson (1979a)
"Ants"	*Polistes* sp.	Carolina	Rabb (1960)
Eciton burchelli	*P. erythrocephalus*	Costa Rica	Young (1979)
Wasps			
Palarus variegatus	*P. nimpha*	Hungary	Moczar (1952)
Palarus humeralis	*P. gallicus*	South Africa	Ferton (1911)
Vespa tropica	*P. hebraeus*	India	Maxwell-Lefroy and Howlett (1909)
Vespa tropica	*Polistes* sp.	Formosa	Sakagami and Fukushima (1957)
Vespa tropica var. *pulcbra*	*P. jadwigae, P. mandarinus*	Japan	Sakagami and Fukushima (1957)
Vespa xanthoptera	*Polistes* sp.	Japan	Sakagami and Fukushima (1957)
Vespa crabro	*Polistes* sp.	Japan	Sakagami and Fukushima (1957)
Vespa crabro	*P. gallicus*	Italy	Turillazzi (pers. observ.)

(continued on next page)

TABLE 2.1 *(continued)*

Predators	Prey	Location	Source
Vespa mandarinia	*P. jadwigae* and other species	Japan	Matsuura and Sakagami (1973)
Vespa tropica	*P. jadwigae* and other species	Japan	Matsuura and Sakagami (1973)
Vespa mongolica	*P. jadwigae* and other species	Japan	Matsuura and Sakagami (1973)
Vespa analis	*P. jadwigae* and other species	Japan	Matsuura and Sakagami (1973)
Vespa crabro	*P. jadwigae* and other species	Japan	Matsuura and Sakagami (1973)
Vespa analis insularis	*P. snelleni, P. chinensis antennalis*	Japan	Matsuura (1973)
	P. jadwigae, P. japonicus		
	P. mandarinus		
Spiders			
Lactodectus mactans	*P. apachus*	California	Gibo and Metcalf (1978)
Unknown species	*Polistes* sp.	Texas	Reed and Vinson (1979a)
Unknown species	*Polistes* sp.	Carolina	Rabb (1960)
Thomisus onustus	*Polistes* sp.	Italy	Guiglia (1972)
Reptiles			
Geckos	*Polistes* sp.	Italy	Pardi (pers. comm.)

Birds

Piranga rubra	*P. pallipes, P. variatus*	Missouri	Rau (1941a)
Cyanocitta cristata	*P. exclamans*	Texas	Pulich (1969)
Pernis apivorus japonicus	"social wasps"	Japan	Sakagami and Fukushima (1957)
Aphelocoma coerulescens	*P. apachus*	California	Gibo and Metcalf (1978)
Mimus polyglottos	*P. apachus* (probably)	California	Gibo and Metcalf (1978)
Passer italiae	*P. gallicus* (probably)	Italy	Turillazzi (pers. observ.)
Piranga rubra	*P. exclamans, P. metricus*	Georgia	Krispyn (1975)
Piranga rubra	*P. annularis*	Georgia	Hermann *et al.* (1975)
Red-winged Blackbirds	*P. fuscatus*	Ontario	Gibo (1978)
Baltimore Orioles	*P. fuscatus*	Ontario	Gibo (1978)
Cardinals	*P. fuscatus*	Ontario	Gibo (1978)
Merops apiaster	"social wasps"	Italy	Guiglia (1972)
Pernis apivorus	"social wasps"	Italy	Guiglia (1972)

Mammals

Cebus variegatus (pet)	*P. versicolor*	Brazil	Zikán (1951)
"Rodent"	*Polistes* sp.	Texas	Reed and Vinson (1979a)
Foxes	*P. biglumis*	Japan	Yamane and Kawamichi (1975)

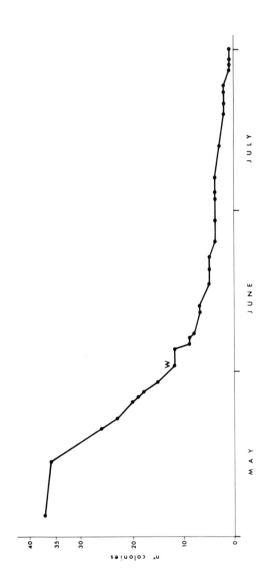

FIGURE 2.1. Survival rate of *P. nimpha* colonies studied near Firenze (Italy) during their active season (all colonies founded by single foundress female). W, emergence of first workers.

Nest Site

Nidification is of fundamental importance to colony success. Implantation on particular substrata diminishes the probability of being discovered by wandering scout ants. Jeanne (1975, 1979) observed that a nest built on a leaf or a secondary branch is relatively more secure than one built on either the trunk or a principal branch, at least in temperate zones. A protected site is also a good requisite against spring coldspells and large predators in temperate regions. In the tropics, where cold is not a problem, even nests implanted in places protected from large predators are exposed to ant attacks. There are very few places that ants cannot reach. *Polistes* nests are common in Guyana in second-growth forests and near clearings where Ecitonini seem to be almost absent (Richards and Richards, 1951). Other social wasps (*Polybia, Mischocyttarus, Synoeca, Stelopolibia*) nest in association with extremely aggressive genera of ants (*Atzeca, Dolichoderus*), which are not their predators and protect them from the attacks of army ants (Richards and Richards, 1951). Oddly enough, *Polistes* seldom forms associations of this kind. Only once has a nest of *P. pacificus* been reported near a large nest of *Azteca cartifex* (Myers, 1935).

Nests associated with other more aggressive social wasps (*P. canadensis* with *Polybia* sp. (Myers, 1935) and *P. carnifex* with *Polybia* sp. in Costa Rica (Corn, 1972)) could be a means of protection against vertebrate predators. The nature of nest associations between social wasps, including *Polistes*, and various species of birds remains obscure (Myers, 1929, 1935). While birds may be protected against vertebrate predators and, above all, from parasite botflies of the genus *Philornis* (Smith, 1968) by their aggressive neighbors, it is less clear why the wasps do not attack the birds. Myers (1935) proposed that the wasps might be warned of the approach of a predator by the crying of the birds or, that the more visible birds nests form "the recognition mark of the ensemble" with the function of danger signals for eventual predators.

Downhower and Wilson (1973) proposed that the presence of Ortoptera katydids (*Ancistocercus inficitus*) near nests of *Polistes* and other social wasps in Costa Rica represents an association similar to those mentioned above. Katydids would function as sentinels against the attacks of vertebrate predators, receiving protection from the wasps in return.

In temperate zones the nest sites of *Polistes* species often differ greatly in a given locality (see Rau, 1931; Nelson, 1968; Reed and Vinson, 1979a, for some North American *Polistes*). In Italy, *P. gallicus* is found almost exclusively on buildings (for example under roof eaves) while *P. foederatus, P. omissus*, and *P. nimpha* nest on shrubs (Pardi, 1942a, and personal observations). Jeanne (1975) thinks that the implantation of stelocyttarous nests on wide surfaces of buildings, very common in tropical and temperate zones, diminishes the probability of being discovered by scout ants and renders predation by ants generally more difficult. This is confirmed by field-study data of the same author (Jeanne, 1979). It has also been suggested that such sites offer better microclimatic conditions to some species (Reed and Vinson, 1979a), although the likelihood of predation by birds and infestation by parasites is enhanced.

In temperate zones, an easily located place, in which a previous nest had been successful could facilitate its rediscovery and reuse, which also increase the probability of pleometrotic nest associations. It is symptomatic that, at least in Italy, species that nest on shrubs are usually haplometrotic while *P. gallicus* commonly founded pleo-metrotic colonies.

Successful nest founding on buildings implies that such sites are concealed from or inaccessible to man. As observed by Reed and Vinson (1979a), the greatest destruction of colonies of three species of *Polistes* in an urban area of Texas was caused by occupants of the houses on which the nests had been built.

Precocious Reproductives Production

Worker production is considerably reduced in the northern zones of distribution in temperate species. In northern Japan the mean ratio of workers:males:foundresses in *P. biglumis* is 2:3:8, while in *P. jadwigae*, from southern Japan, it is 8:1:1 (Yamane, 1969). This is evidently more an adaptation to the shortest season of nidification than a response to predator pressure. However, short colonial cycles were also found in the tropical genus *Mischocyttarus*, where worker production is reduced in favor of reproductive production (Jeanne, 1975). The production of reproductives in tropical *P. canadensis* occurs comparatively early in the cycle in contrast to North American *P. fuscatus* (West, 1969). The same occurs in *P. tepidus malayanus* of New Guinea, a species that uses the same cell in raising two consecutive larvae (Yamane and Okazawa, 1977).

Nonworker females in tropical species that have no winter ovarian diapause can found a new nest even when the maternal nest is still active.

Refounding of Destroyed Colonies and Protraction of the
Reproductive Capacity

In temperate *Polistes*, nests are easily refounded if destruction occurs at the beginning of the colony cycle, without compromising their success. During this stage, active defense of the nest by the foundresses is very mild and they will leave the nest even if slightly disturbed. If nest destruction occurs later in the season, before emergence of the workers, then success is compromised because too little time remains for a new colony to produce reproductives.

Although there are no impediments to a perennial colony cycle in tropical *P. canadensis* in Colombia, where the queen is superseded by a daughter in colony leadership, colonies seldom last more than one year (West, 1969). The maternal nest is abandoned and new colonies are established nearby (the largest nest reported had 750 cells). These wasps, which have no defense against driver ants, could find advantages in dividing their brood among little consecutive colonies and limiting the size of their colonies even when conditions are extremely favorable.

Mimetic Nidification

Nest shape in some *Polistes* species is highly mimetic. The cell walls in nests of *P. goeldii* (West, 1969) are only partially shared, and the general elongated shape resembles a twig. The same kind of nest is built by *P. stenopus* in the Solomon Islands (van der Vecht, 1971, quoted by Yamane and Okazawa, 1977) and by *Ropalidia, Mischocyttarus punctatus*, and some Stenogastrinae. According to Jeanne (1975) and West (1969), these nests represent primitive stages, but it is also true that they are very mimetic with vegetation (Yamane and Okazawa, 1977). Moreover, the wasps occasionally collect nest building material from the substrate on which the nest is implanted, rendering the color of the construction very similar to its support.

Polistes nimpha generally builds its nests 3–15 cm above the ground on shrubs that are quickly concealed by grass.

Direct Defense against Predators

Vertebrate predators. Just as every other social wasp, *Polistes* protect themselves directly from vertebrate predators by their sting. *Polistes* can be considered as moderately aggressive wasps. *Polistes* have no sting autotomy, in contrast to other polistine wasps such as *Polybia,*

Synoeca, and some *Brachygastra* (Hermann, 1971), but their sting is deadly for small animals and can cause severe allergic reactions in sensitive individuals.

A lone wasp rarely attacks if molested; only if it is in extreme danger or if the colony is threatened will it behave aggressively. While *Polistes* wasps seldom use their sting when hunting, they do use it against conspecifics, such as in fights between foundresses in pleo-metrotic spring associations (Pardi, 1946).

The bright characteristic coloration of many species is an aposematic signal to potential predators, as in many other venomous animals. Nonvenomous insects are known to imitate the fundamental vespoid coloration of light (yellow or brown) and dark stripes (Wickler, 1976). Even *Polistes* males show the same coloring as the females and sometimes (*P. canadensis*, West, 1969) there is almost no sexual dimorphism. If grasped by a predator the males try to "sting" like the females.

When an intruder approaches the colony the females rise and spread their wings and antennae, facing the intruder with bodies raised from the nest. If the stimulus is intense they begin buzzing their wings and eventually attack, returning at once to the nest (Spieth, 1948; West, 1969; Corn, 1972; C.K. Starr, unpublished observations). Spieth (1948) observed that the first wasps to react are those in the upper part of the nest. Vibrations of the substratum (the paper nest) caused by swift or jerking movements of the wasps are important in communicating alarm to the nest mates (West, 1969). The posture assumed by the wasps may represent a warning mech-anism (Hermann and Dirks, 1975), while the contemporaneous response of many individuals suddenly increases the size of the colony (Figure 2.2). Starr (1980) observed that even the males, when on nest, may threaten like the females. Flying attacks are usually directed toward moving targets (which explains why man is usually stung on the hands and face) and are led almost exclusively by workers.

Yamane (1969) observed that even young foundresses of *P. snelleni* can attack, which may be related to their abundance in the nest population. C. K. Starr (unpublished observations) recently quanti-fied the behavioral responses of some American *Polistes* to external stimulations. Though Rau (1939a) reports that some *Polistes* (for example *P. fuscatus variatus*) occasionally emit a very "pleasantly sweetish odor" when disturbed, according to Maschwitz (1964) *Polistes* do not possess alarm pheromones.

Wasps can carry on their vigilance even at some distance from the nest. If this is situated in a closed space, some sentinels (according to Nelson, 1968, in *P. carolinus=P. rubiginosus*) may stay at the entrance as happens in the ground colonies of Vespinae (Spradbery, 1973).

FIGURE 2.2. Reaction of *P. gallicus* colony to external stimulus (sudden movement of hand in proximity of nest).

Defense against ants. In contrast to a wide junction with the sub-stratum the narrow stem of stelocyttarous nests in *Polistes* (and other genera of Polistinae) "would lessen the probability of a wandering predator discovering the exposed brood" (West, 1969).

Jeanne (1970) discovered that female *Mischocyttarus drewseni* coat the nest stem with an ant repellent that they secrete and apply during a behavior called "rubbing." A similar behavior was observed in *Belonogaster griseus* (Marino Piccioli and Pardi, 1970), in *Polistes carnifex* (Corn, 1972), in various species of North American *Polistes* (Hermann and Dirks, 1974), in *Ropalidia cincta* (Darchen, 1976), in three species of European *Polistes* (Turillazzi and Ugolini, 1978, 1979) (Figure 2.3) and in *P. tenebricosus* from Bali (S. Turillazzi and Pardi, unpublished observations). The substance rubbed on the stem and external part of the nest is secreted by tegumental unicellular glands located near a lightly sclerotized area, with a tiny brush of hairs at the anterior edge of the terminal gastral sternite in the females. The structure, called van der Vecht's organ, is present in many species of Vespidae (van der Vecht, 1968) and is particularly developed in *Polistes* (Figure 2.4). Clusters of tegumental glands are also found at the anterior edge of the remaining gastral sternites (Turillazzi, 1979), particularly the sixth (Bordas, 1908; Hermann and Dirks, 1974; Turillazzi, 1979; Delfino *et al.*, 1979; Post and Jeanne, 1980). The van der Vecht's organ clearly acts as a brush in spreading the secretion. Foundresses of *P. gallicus, P. foederatus, P. nimpha* (Turillazzi and Ugolini, 1979) and *P. annularis* (Hermann and Dirks, 1974*); the frequency of this behavior diminishes per colony and per wasp after appearance of the workers in *P. nimpha* (Figure 2.5) and *P. gallicus* (Turillazzi and Ugolini, 1979) while it is still often practiced on postemergence colonies of *P. fuscatus* (Post, 1980). The stem is frequently rubbed in the early morning and evening and hardly at all near noon (Figure 2.6). The wasps rub the pedicel after careful inspection, after the pedicel's enlargement and usually before off-the-nest activities. The presence of an ant near the nest sparks a behavioral sequence in adult *Polistes* (West, 1969; Hermann and Dirks, 1974; Turillazzi and Ugolini, 1979; Post, 1980) typified by the reaction of *P. fuscatus*: the wasps "stand buzzing their wings near the pedicel . . . and peck sharply at the ant with their mandibles" (West, 1969). A similar action occurs in three species of *Polybia* and one of *Apoica* (Jeanne, 1975) and in *Parischnogaster nigricans* (Turillazzi and Pardi, 1982). When a *P. gallicus* nest is approached by a worker of *Crematogaster scutellaris*, the females begin to rub the stem within 1 minute of their detection of the

---

*Rub the nest stem particularly during the pre-emergence period.

ant (Figure 2.7) (Turillazzi and Ugolini, 1979). Similar reactions were prompted even by small amounts of chemical irritants (formic acid, acetic acid) placed near the nest. Anything touching the nest, such as grass stalks, leaves, or sticks placed experimentally, are rubbed by the wasps after attempts to remove them fail. The substance is mildly repellent to ants even in *Polistes* (Hermann and Dirks, 1974; Turillazzi and Ugolini, 1979; Post, 1980). Experiments have thus far been made only on temperate species, and it would be interesting to know if tropical species (under greater threat of ant predation) produce a more powerful substance. Post and Jeanne (1980) found recently that the tropical species *P. canadensis* has more numerous and larger glands in the terminal gastral sternite than the temperate species *P. fuscatus*. It is probable, but not yet certain, that rubbing has the same significance in *Ropalidia* and *Belonogaster*. Other social wasps, such as *Nectarinella championi* (Schremmer, 1977) and some species of Stenogastrinae (Turillazzi and Pardi, 1980) secrete ant-repellent substances.

Chemical defense may be extremely useful during the early stages of colony development especially in haplometrotic nests. In fact, the nest is completely defenseless when the foundress is off it. Once the first workers emerge the colony has a means of continuous active defense. Some workers almost always guard the nest stem, while a lone foundress takes up this position at night (Hermann and Dirks, 1974; Turillazzi and Ugolini, 1979; Post, 1980). As soon as an ant is detected it is grasped and thrown to the ground.

Chemical defense of the nest is useless against the attack of army ants, which are not uncommon in the tropics (Jeanne, 1975; Young, 1979). Raids by *Eciton burchelli* have a catastrophic effect on colonies of *P. erythrocephalus*, completely destroying the brood and causing the adult wasps to abandon the nests for several days (Young, 1979). Even adults of neighboring but untouched colonies evacuate their nests while the raid is on.

New Colony Establishment and Defense

Social wasps form their colonies by swarming (nonworker females, workers, and males found a new colony together), pleo-metrotic foundation, or haplometrotic foundation. Whether all three methods are present in *Polistes* has not been established, mainly because a morphological and thus irreversible castal differentiation has not yet been demonstrated in the genus. According to Weyrauch (1939, p. 178) *P. gallicus* should present true swarming in the Sahara oasis; however, his observations have not yet been confirmed. On the other hand, the latitudinal gradient that seems to exist in colony

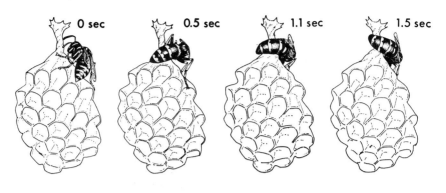

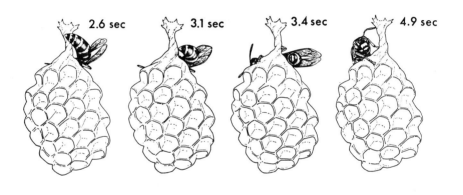

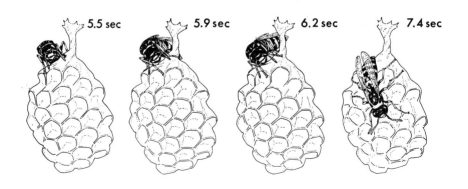

foundation method in this European species (with pleometrosis more common in Mediterranean regions and haplometrosis as a rule in northern countries) has not been found in North American *Polistes*.

Pardi (1942b, 1946) and later many other authors, pointed out the presence of a domination hierarchy in females belonging to pleometrotic associations. Only the alfa female exhibits a queen-like behavior and egg-laying capacity, while the other females (auxiliaries) undergo ovarial regression and assume worker-like behavior. Females in these associations are closely related and are often sisters (West, 1969; Klahn, 1979). The advantages presented by multiple- as opposed to single-foundress colonies of *P. fuscatus* in the presence of predation (Gibo, 1978) is one of the evolutionary aspects of this phenomenon (see also Hamilton, 1964; West, 1967a; West Eberhard 1975, 1978; Starr, 1979). Gibo (1978) discovered that during the pre-emergence period such predators as birds are capable of destroying both haplometrotic and pleometrotic colonies. Several foundresses have the same defense power as a single foundress. However the occupants of destroyed multiple-foundress colonies rebuild significantly more than those of single-foundress colonies. Moreover, when the predation level is high, multiple-foundress colonies are more productive per colony and per foundress. It seems clear that both environmental conditions and predation pressure play a role in the evolution of pleometrotic associations though, as Gibo observes, this phenomenon has not yet been completely understood. The disappearance of foundresses during the pre-emergence period is fatal for haplometrotic colonies, whereas in pleometrotic ones the cycle continues because of the remaining females.

Gamboa (1978) believes that where nest density is high the association of females of *P. metricus*, probably sisters, favors defense against usurpation from conspecific but unrelated females rather than against predators or parasites. Where the wasp density is high, single-foundress nests are more subjected to nest usurpation than multiple-foundress nests. Under these conditions it would be advantageous for the less dominant females, in terms of longevity and colony productivity, to "join an established queen rather than to attempt to construct a nest."

Colony Odor

In *Polistes*, as for most other social insects (Wilson, 1972), recognition of each colony member is probably olfactory (Rau, 1939b;

FIGURE 2.3. Rubbing movements in *P. nimpha* (drawn from video recording).

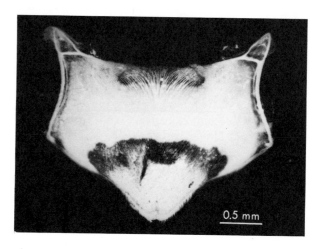

FIGURE 2.4.   Terminal gastral sternite of female *P. gallicus* showing van der Vecht's organ at the anterior edge.

Pardi, 1947). Strangers, even if conspecific, are attacked as soon as they approach a new colony. According to West (1969), the approaching behavior to the nest and dominance relations among nest residents play a role in nest mates recognition as well. Resident wasps usually respond to nonresident ones who approach the nest in an aggressive fashion, waving the front legs and buzzing their wings and sometimes attacking in flight. Also, the possibility of individual recognition by learned visible characteristics cannot be excluded. Nest mates recognition may be important to prevent usurpations from more dominant females and predation of brood by wasps belonging to other colonies.

SOCIAL PARASITISM

Sulcopolistes, characterized by almost perfect resemblance to *Polistes*, lack a worker caste, and the females are social parasites of European *Polistes* colonies. The parasite female attacks the host colony in the late pre-emergence period, subjugating the host queen, which is seldom killed or driven away, and behaving as the dominant female in a pleometrotic association (Scheven, 1958). The usurper wins the fight because she is larger and stronger than the *Polistes* female. The workers of *Polistes* that emerge later then rear the brood of the *Sulcopolistes* female.

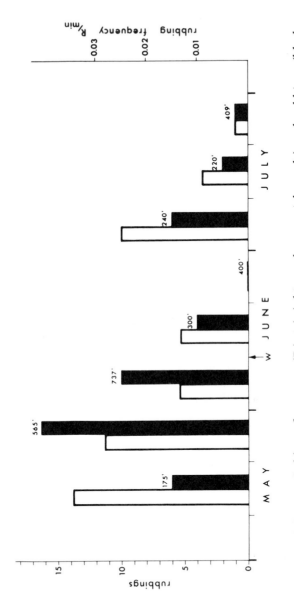

FIGURE 2.5. Rubbing frequency (R/min) (white columns, right scale) and rubbings (black columns, left scale) observed on several *P. nimpha* colonies during active season. Numbers on columns indicate minutes of observation.

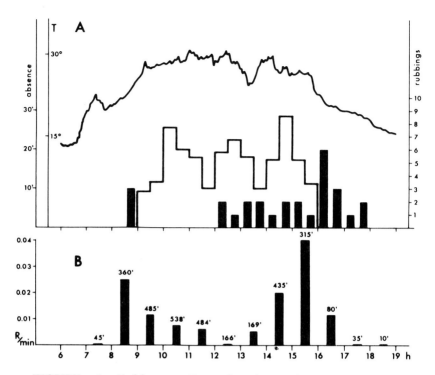

FIGURE 2.6. Rubbings in *P. nimpha* colonies during day. (A) Over 12-hour period on single foundress pre-emergence nest in field on May 28, 1980. Black columns: number of rubbings performed every 30 min (right scale). White columns: time spent in off-the-nest activities (left scale). Lines indicate temperature. (B) Frequency of rubbing (R/min) at various times of day on several colonies during active season. Numbers on columns indicate minutes of observation.

I do not know any particular defensive mechanism used by *Polistes* against their social parasites. As parasitism generally concerns only haplometrotic colonies (Scheven, 1958), it suggests that pleometrotic colonies are more protected from *Sulcopolistes* attacks.

PARASITISM

Parasites and Parasitoids Attacking *Polistes*

Numerous parasites, parasitoids, and symbionts are associated with colonies of *Polistes* wasps. Nelson (1968) reports 42 species

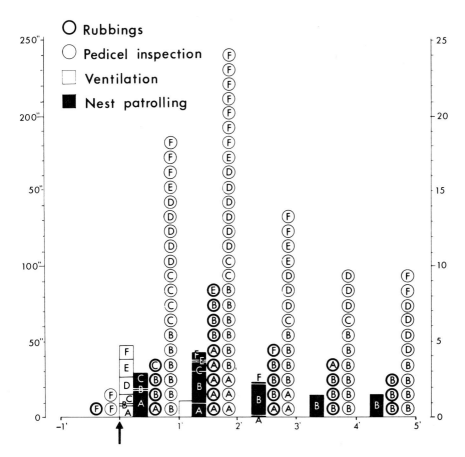

FIGURE 2.7. Behavior observed on *P. gallicus* colonies from 1 minute before to 5 minutes after having placed dead *Crematogaster scutellaris* ant 1–2 cm from nest (A–C) Bigynic nests; (D–F) monogynic nests. *Left scale*: seconds that each wasp devoted to ventilation and nest patrolling; *right scale*: total number of times each wasp performed rubbing and pedicel inspection.

representing 28 families of arthropods taken from the nest of some North American *Polistes* (*P. annularis, P. exclamans, P. metricus*). Information on European species can be found in Guiglia (1972) and on South American, Australian, and African species in Vesey-Fitzgerald (1938), Richards and Richards (1951), and Richards (1969, 1978). Parasitism appears to be particularly frequent in the New World. Ballou (1934, quoted by Spradbery, 1973) states that the moth *Chacoela* (=*Dicymolomia*) sp. has eliminated *Polistes* from some islands in the West Indies. Infestations of *Chalcoela pegasalis* (=*iphitalis*) (Rau, 1930b, 1941b;

Gillaspy, 1971, 1973; Rabb, 1960; Starr, 1976, 1978), *Elasmus polistis* (Ichneumonidae) (Reed and Vinson, 1979b) and *Pachysomoides* sp. (West, 1969) are particularly frequent in the United States. In some localities of Kansas, 73.7% of *P. fuscatus* and *P. metricus* colonies examined are parasitized by *Chalcoela* and 63.4% of *P. annularis* colonies in Georgia (Starr, 1978). The larvae of *Chalcoela* usually feed on the larval fecal pellets of *Polistes* but can perforate the cell walls and devour the host pupae. Liquid droplets regurgitated by wasp larvae and normally taken by adult members of the colony were observed to be taken by moth larvae in a clear case of cleptotrophallaxis (Gillaspy, 1973). Nest structures are also weakened by the internal tunnels excavated by *Chalcoela* larvae. Parasitism is seldom reported in European or Japanese species of *Polistes* and never at the levels found in American species. Infestations of Lepidoptera are rare in Europe, though *P. gallicus* is often parasitized by the ichneumonid *Endurus argiolus*.

Defense Mechanisms Against Parasites

The dispersal of nest foundations observed by West (1967b, 1968) in *P. exclamans* may represent "a response to the pressure of parasitism, since the incidence of chalcidoid parasitoids and of lepidopterous and dipterous nest associates was noticeably lower among the *P. exclamans* nests than among those of *P. metricus* collected the same fall" (Eickwort, 1969). The presence of *Chalcoela* could limit nest reutilization in *Polistes*, although other factors may be involved (Starr, 1976, 1978).

In the lower Amazon region of Brazil, *P. canadensis canadensis* builds nests consisting of several combs, independently attached, with each comb usually containing an average of thirty cells. The wasps do not reuse the cells for rearing a second batch of brood when these have been infested by an undescribed tineid moth, and chew down part of the cell walls. It appears that the wasps occasionally clip the peduncles of several combs whether these are infested or not (Jeanne, 1979). Jeanne hypothesizes that this behavior could reduce brood loss due to the moths. The distribution of brood in many isolated combs, though involving a greater amount of nesting material and reduction in reared larvae (the author calculates that *P. canadensis* could construct a single comb of 336 cells using the same amount of material it usually uses to build ten separate combs of 30 cells each), reduces the probability of attack by moths on each little batch of brood even in the vicinity of infested combs. The risk would be greater if the brood were collected in a single-comb nest. In all colonies examined the risk of infestation was calculated to be 30%, as compared to the 56% value for single-comb nests. The "chewing down of the cell walls and subsequent removal or coating of the meconium may represent an attempt by adult wasps to remove or kill moth eggs or larvae" (Jeanne, 1979). Even nest clipping reduces the incidence of parasitism.

Other species of *Polistes* build comb aggregations: *P. aterrimus* and *P. carnifex* (Rodriguez, 1968) and *P. infuscatus* in Surinam (Jeanne, 1979). The destruction of cell walls in some North American *Polistes* could be a means of dealing with the larvae of *Chalcoela* and, perhaps, the fly *Sarcophaga polistensis* (Gillaspy, 1973).

Direct attacks by the wasps against parasites are infrequent. Adult *P. major* have been seen to drive adult *Chalcoela* away from the nest (Gillaspy, 1973). As soon as female *P. fuscatus* detect the ichneumonid *Pachysomoides fulvus* they dart forward and chase her away, and then quickly begin to patrol the edge of the nest, darting and flipping their wings ("parasite alarm") (West, 1969). This could induce the other females to behave similarly. West never observed such behavior wthout finding a *Pachysomoides* female near the nest. Female *P. canadensis* behave similarly in the presence of *Pachysomoides stupidus* and an unidentified tachinid fly parasite (West, 1969). Female *P. gallicus* reared in captivity react mildly to the ichneumonid *Endurus argiolus* (personal observations). As soon as the wasps detect the parasite they attack it in flight, grasping it with their mandibles and throwing it to the ground, but show no particular behavior on returning to the nest.

Active defense against parasites is much less common than defense against predators and seems to be less effective. However, the evolution of mechanisms and behavior limiting parasite infestations are as yet little known.

REFERENCES

Ballou, H. A. (1934). Notes on some insect pests in the Lesser Antilles. *Trop. Agric. Trin.* 11, 210–212.

Bordas, L. (1908). Les glandes cutanées de quelques Vespides. *Bull. Soc. Zool. Fr.* 33, 59–64.

Corn, M. L. (1972). Notes on the biology of *Polistes carnifex* (Hymenoptera, Vespidae) in Costa Rica and Colombia. *Psyche* 79 (3), 150–157.

Darchen, R. (1976). *Ropalidia cincta*, guêpe sociale de la savane de Lampto (Cote-d'Ivoire) (Hym. Vespidae). *Annls. Soc. Ent. Fr.* (N.S.) 12 (4): 579–601.

Delfino, G., M. T. Marino Piccioli, and C. Calloni. (1979). Fine structure of the glands of Van der Vecht's organ in *Polistes gallicus* (L.) (Hymenoptera Vespidae). *Monit. Zool. Ital.* (NS) 13, 221–247.

Downhower, J. F., and D. E. Wilson. (1973). Wasps as a defense mechanism of katydids. *Am. Midl. Nat.* 89 (2), 451–455.

Eickwort, K. R. (1969). Separation of the castes of *Polistes exclamans* and notes on its biology (Hym. Vespidae). *Insectes Soc.* 16 (1), 67–72.

Ferton, C. (1911). Notes detachées sur l'instinct des Hymenoptères mellifères et ravisseurs. *Ann. Soc. Entomol. Fr.* 80, 351–412.

Fiebrig, K. (1907). Eine Wespen zerstörende Ameise aus Paraguay, *Eciton vagans* Olivier. *Z. Wiss. Insekt. Biol.* 3, 83–87.

Gamboa, G. J. (1978). Intraspecific defense: Advantage of social cooperation among paper wasp foundresses. *Science* 199, 1463–1465.

Gibo, D. L. (1978). The selective advantage of foundress associations in *Polistes fuscatus* (Hymenoptera: Vespidae): A field study of the effects of predation on productivity. *Can Entomol.* 110, 519–540.

Gibo, D. L., and R. A. Metcalf. (1978). Early survival of *Polistes apachus* (Hymenoptera: Vespidae) colonies in California: A field study of an introduced species. *Can Entomol.* 110: 1339–1343.

Gillaspy, J. E. (1971). Papernest wasps (*Polistes*): Observations and study methods. *Ann. Entomol. Soc. Am.* 64 (6): 1357–1361.

Gillaspy, J. E. (1973). Behavioral observations on paper-nest wasps (genus *Polistes*; family Vespidae; order Hymenoptera). *Am. Midl. Nat.* 90 (1), 1–12.

Guiglia, D. (1972). Les Guêpes sociales (Hymenoptera Vespidae) d'Europe occidentale et septentrionale. Masson et Cie, Paris.

Hamilton, W. D. (1964). The genetical evolution of social behavior, I and II. *J. Theor. Biol.* 7, 1–52.

Hermann, H. R. (1971). Sting autotomy, a defensive mechanism in certain social Hymenoptera. *Insectes Soc.* 18, 111–120.

Hermann, H. R., R. Barron, and L. Dalton (1975). Spring behavior of *Polistes exclamans* (Hymenoptera:Vespidae:Polistinae). *Entomol. News* 86, 173–178.

Hermann, H. R., and T. F. Dirks. (1974). Sternal glands in polistine wasps:Morphology and associated behavior. *J. Ga. Entomol. Soc.* 9 (1), 1–8.

Hermann, H. R., and T. F. Dirks. (1975). Biology of *Polistes annularis* (Hymenoptera Vespidae). I. Spring behavior *Psyche* 82 (1), 97–107.

Jeanne, R. L. (1970). Chemical defense of brood by a social wasp. *Science* 168, 1465–1466.

Jeanne, R. L. (1972). Social biology of the neotropical wasp *Mischocyttarus drewseni*. *Bull. Mus. Comp. Zool.* 144 (3), 63–150.

Jeanne, R. L. (1975). The adaptiveness of social wasp nest architecture. *Q. Rev. Biol.* 50, 267–287.

Jeanne, R. L. (1979). Construction and utilization of multiple combs in *Polistes canadensis* in relation to the biology of a predaceous moth. *Behav. Ecol. Sociobiol.* 4, 293–310.

Jeanne, R. L. (1979). A latitudinal gradient in rates of ant predation. *Ecology* 60 (6), 1211–1224.

Klahn, J. E. (1979). Philopatric and nonphilopatric foundress associations in the social wasp *Polistes fuscatus*. *Behav. Ecol. Sociobiol.* 5, 417–424.

Krispyn, J. W. (1975). Paper making wasps. An annotated key to the social Vespidae of Georgia. Thesis, Univ. Georgia, 202 pp.

Marino Piccioli, M. T., and L. Pardi. (1970). Studi sulla biologia di *Belonogaster* (Hymenoptera Vespidae). 1. Sull'etogramma di *Belonogaster griseus* (Fab.). *Monit. Zool. Ital.* (NS) Suppl. 3: 63–150.

Maschwitz, U. (1964). Gefahrenalarmstoffe und Gefahrenalarmierung bei sozialen Hymenopteren. *Z. Ver. Physiol.* 47 (6), 596–655.

Matsuura, M. (1973). Intracolonial polyethism in *Vespa*. III. Foraging Activities. Life Study (Fukui) 17 (3–4): 81–99.

Matsuura, M. and S. F. Sakagami. (1973). A bionomic sketch of the giant hornet, *Vespa mandarinia*, a serious pest for Japanese Apiculture. *J. Fac. Sci. Hokkaido Univ. (Ser. 6) Zool.* 19 (1), 125–162.

Maxwell-Lefroy, H. and F. M. Howlett. (1909). Indian insect life. A manual of the insects of the Plains (Tropical India). Thacker, Spink and Co., Calcutta.

Moczar, L. (1952). Contribution á l'ethologie du *Palarus variegatus* F. (Hym.) *Ann. Hist. Nat. Mus. Nac. Hungarici*, (NS) 2, 119–124.

Myers, J. G. (1929). The nesting together of birds, wasps and ants. *Proc. Entomol. Soc. London.* 4, 80–90.

Myers, J. G. (1935). Nesting associations of birds with social insects. *Trans. R. Entomol. Soc. London* 83, 11–22, 1 plate.

Nelson, J. M. (1968). Parasites and symbionts of nests of *Polistes* wasps. *Ann. Entomol. Soc. Am.* 61, 1528–1539.

Pardi, L. (1942a). Ricerche sui Polistini. IV. Note critiche sulla nidificazione di *Polistes gallicus* (L.) e di *Polistula bischoffi* Weyrauch. *Atti Soc. Toscana Scienze Nat.* 51, 3–13.

Pardi, L. (1942b). Ricerche sui Polistini. V. La poliginia iniziale in *Polistes gallicus* (L.). *Boll. Ist. Entomol. Univ. Bologna* 14, 1–106.

Pardi, L. (1946). Ricerche sui Polistini. VII. La dominazione e il ciclo ovarico annuale in *Polistes gallicus* (L.). *Boll. Ist. Ent. Univ. Bologna* 15, 25–84.

Pardi, L. (1947). Beobachtungen über das Interindividuelle Verhalten bei *Polistes gallicus* (Untersuchungen über die Polistini No. 10). *Behaviour* 1, 138–172.

Post, D. C. (1980). Chemical defense in the temperate social wasp, *Polistes fuscatus* (Hymenoptera; Vespidae). M.Sc. Thesis, Univ. Wisconsin-Madison, 109 pp.

Post, D. C., and R. L. Jeanne. (1980). Morphology of the sternal glands of *Polistes fuscatus* and *P. canadensis* (Hymenoptera: Vespidae). *Psyche* 87: 49–58.

Pulich, W. M. (1969). Unusual feeding behavior of three species of birds. *Wilson Bull.* 81, 472.

Rabb, R. L. (1960). Biological studies of *Polistes* in North Carolina (Hymenoptera: Vespidae). *Ann. Entomol. Soc. Am.* 53, 111–121.

Rau, P. (1930a). Ecological and behavior notes on the wasp *Polistes pallipes*. *Can. Entomol.* 62, 143–147.

Rau, P. (1930b). Life history notes on the wasps, *Polistes annularis. Can. Entomol.* 62 (6), 119–120.

Rau, P. (1931). The nest and nesting sites of four species of *Polistes* wasps. *Bull.*

Brooklyn Entomol. Soc. 26 (3): 111–118.

Rau, P. (1939a). Studies in the ecology and behavior of *Polistes* wasps. *Bull. Brooklyn Entomol. Soc.* 34, 36–44.

Rau, P. (1939b). The instinct of animosity and tolerance in queen *Polistes* wasps. *J. Comp. Psychol.* 27, 259–269.

Rau, P. (1941a). Birds as enemies of *Polistes* wasps. *Can. Entomol.* 73, 196.

Rau, P. (1941b). Observations on certain lepidopterous and hymenopterous parasites of *Polistes* wasps. *Ann. Entomol. Soc. Am.* 34, 355–366.

Reed, H. C., and S. B. Vinson. (1979a). Nesting ecology of paper wasps (*Polistes*) in a Texas urban area (Hymenoptera: Vespidae) *J. Kans. Entomol. Soc.* 52 (4), 673–689.

Reed, H. C., and S. B. Vinson. (1979b). Observations of the life history and behavior of *Elasmus polistis* Burks (Hymenoptera:Chalcidoidea:Eulophidae). *J. Kans. Entomol. Soc.* 52 (2), 247–257.

Richards, O. W. (1969). The biology of some W. African social wasps (Hymenoptera:Vespidae: Polistinae). *Mem. Soc. Entomol. Ital.* 48, 79–93.

Richards, O. W. (1978). The Australian social wasps (Hymenoptera:Vespidae). *Austral. J. Zool.* (Suppl. ser.) 61, 1–132.

Richards, O. W., and M. J. Richards. (1951). Observations on the social wasps of South America (Hymenoptera Vespidae). *Trans. R. Entomol. Soc. London* 102, 1–170.

Rodriguez, V. M. (1968). Estudo sôbre vespas sociãs do Brasil (Hymenoptera Vespidae). Doctoral thesis Fac. Filos. Cienc. Letras de Rio Claro, Universidade de Campinas.

Sakagami, S. F., and K. Fukushima. (1957). Some biological observation on a hornet, *Vespa tropica* var. *pulchra* (Du Buysson), with special reference to its dependence on *Polistes* wasps (Hymenoptera). *Treubia* 24 (1), 73–82.

Scheven, J. (1958). Beitrag zur Biologie der Schmarotzefelden-wespen. *Insectes Soc.* 5 (4), 409–438.

Schremmer, F. (1977). Das Baumrinden-Nest der neotropischen Faltenwespe *Nectarinella championi*, umgeben von einem Leimring als Ameisen-Abwehr (Hymenoptera: Vespidae). *Entomol. Germ.* 3 (4), 344–355.

Smith, N. G. (1968). The advantage of being parasitized. *Nature* 219, 690–694.

Spieth, H. T. (1948). Notes on a colony of *Polistes fuscatus hunteri* Bequaert. *N. Y. Entomol. Soc.* 56, 155–169.

Spradbery, J. P. (1973). *Wasps*, 14 ed. Sidwick & Jackson, London. XIV +

Starr, C. K. (1976). Nest reutilization by *Polistes metricus* (Hymenoptera: Vespidae) and possible limitation of multiple foundress associations by parasitoids. *J. Kans. Entomol. Soc.* 49 (1), 142–144.

Starr, C. K. (1978). Nest reutilization in North American *Polistes* (Hymenoptera:Vespidae):Two possible selective factors. *J. Kans. Entomol. Soc.* 51 (3), 394–397.

Starr, C. K. (1979). Origin and evolution of insect sociality: A review of modern theory. In *Social Insects* (H. R. Hermann, ed.) Vol. 1, pp. 35–79. Academic Press, New York.

Starr, C. K. (1980). Interspecific differences in defensive tactics of *Polistes* wasps (Vespidae). Abstracts XVI International Congress of Entomology. Kyoto, Japan, Aug. 3–9, 1980, p. 430.

Suzuki, T. (1978). Area, efficiency and time of foraging in *Polistes chinensis antennalis* Pérez (Hymenoptera, Vespidae). *Jap. J. Ecol.* 28, 179–189.

Turillazzi, S. (1979). Tegumental glands in the abdomen of some European *Polistes* (Hymenoptera, Vespidae). *Monit. Zool. Ital.* (NS) 13, 67–70.

Turillazzi, S., and R. Cervo. (in preparation). Biology and social behavior of *Polistes nimpha* (Christ)(Hymenoptera Vespidae).

Turillazzi, S., and L. Pardi. (1981). Ant guards on nests of *Parischnogaster nigricans* (Cameron) subsp. *serrei* (Buysson) (Stenogastrinae). *Monit. Zool. Ital. (NS) (Suppl.)* 15:1–7.

Turillazzi, S., and L. Pardi. (1982). Changes in the social behavior of *Parischnogaster nigricans* (Cameron) subsp. *serrei* (Buysson) (Hymenoptera: Vespidae in Java). *Ann. Entomol. Soc. Amer.* 75:657–664.

Turillazzi, S., and A. Ugolini. (1978). Nest defense in European *Polistes* (Hymenoptera Vespidae). *Monit. Zool. Ital.* (NS) 12, 72.

Turillazzi, S. and Ugolini, A. (1979). Rubbing behavior in some European *Polistes* (Hymenoptera Vespidae). *Monit. Zool. Ital.* (NS) 13, 129–142.

van der Vecht, J. (1968). The terminal gastral sternite of female and worker social wasps (Hymenoptera Vespidae). *Proc. K. Ned. Akad. Wet. (C)* 71 (4), 411–422.

van der Vecht, J. (1971). The subgenera *Megapolistes* and *Stenopolistes* in the Solomon Islands (Hymenoptera, Vespidae:*Polistes* Latreille). Entomological essays to commemorate the retirement of Prof. K. Yasumatsu, pp. 87–106. Fukuoka.

Vesey-Fitzgerald, D. (1938). Social wasps (Hym. Vespidae) from Trinidad, with a note on the genus *Trypoxylon* Latreille. *Trans. R. Entomol. Soc. London* 87, 181–191.

West, M. J. (1967a). Foundress associations in Polistine wasps: dominance hierarchies and the evolution of social behavior. *Science* 1957, 1584–1585.

West, M. J. (1967b). The social biology of polistine wasps. Univ. Microfilms. Ph.D. thesis, University of Michigan, 220 p. (quoted by Eickwort, 1969).

West, M. J. (1968). Range extension and solitary nest founding in *Polistes exclamans* (Hymenoptera:Vespidae). *Psyche*, 75 (2), 118–123.

West, M. J. (1969). The social biology of Polistine wasps. *Misc. Publs. Mus. Zool. Univ. Mich.* 140, 1–101.

West Eberhard, M. J. (1975). The evolution of social behavior by kin selection. *Quart. Rev. Biol.* 50, 1–33.

West Eberhard, M. J. (1978). Poligyny and the evolution of social behavior in wasps. *J. Kansas Ent. Soc.* 51 (4), 832–856.

Weyrauch, W. (1939). Zur Systematik der Palearktischen Polistinen auf biologischer Grundlage. *Archiv. f. Naturgesch. n. F.* 8 (2), 145–197.

Wickler, W. (1976). Mimicry and camouflage. In Grzimek's Encyclopedia of Ecology, pp. 134–155. (Grzimek, ed.) Van Nostrand Reinhold, New York.

Wilson, E. O. (1972). *The Insect Societies.* Belknap Press of Harvard Univ. Press. Cambridge, Mass.

Yamane, S. (1969). Preliminary observations on the life history of two Polistine wasps, *Polistes snelleni* and *P. biglumis* in Sapporo, northern Japan. *J. Fac. Sci. Hokkaido Univ., Ser. VI Zool.*, 17, 78–105.

Yamane, S., and T. Kawamichi (1975). Bionomic comparison of *Polistes biglumis* (Hymenoptera, Vespidae) at two different localities in Hokkaido, northern Japan with reference to its probable adaptation to cold climate. *Kontyû, Tokyo*, 43, 214–232.

Yamane, S. and T. Okazawa (1977). Some biological observations on a paper wasp, *Polistes (Megapolistes) tepidus malayanus* Cameron (Hymenoptera, Vespidae) in New Guinea. *Kontyû, Tokyo*, 45 (2), 283–299.

Yoshikawa, K. (1962). Introductory studies on the life economy of Polistine wasps. VI. Geographical distribution and its ecological significances. *J. Biol. Osaka City Univ.* 13, 19–43.

Young, A. M. (1979). Attacks by the army ant *Eciton burchelli* on nests of the social paper wasp *Polistes erythrocephalus* in Northeastern Costa Rica. *J. Kansas Entomol. Soc.* 52 (4), 759–768.

Zikan, J. F. (1951). Polymorphismus und ethologie der sozialen Faltenwespen (Vespidae, Diploptera). *Acta Zool. Lilloana*, 11, 5–51.

3
Vespine Defense
Roger D. Akre
and
Hal C. Reed

INTRODUCTION

The subfamily Vespinae is comprised of the hornets (*Provespa* Ashmead, *Vespa* L.) and their smaller relatives, the yellowjackets (*Dolichovespula* Rohwer, and *Vespula* Thomson) (Akre and Davis, 1978). *Provespa* occurs only in the tropical areas of the Oriental region, while *Vespa* has both tropical and temperate species; however, most occur in eastern Asia. *Dolichovespula* and *Vespula* are primarily temperate in distribution, with the possible exception of *Vespula squamosa* (Drury) (Miller, 1961).

Due to the paucity of published information on defensive behaviors of species of *Provespa* and *Vespa*, most of the material in this chapter discusses defense by species of *Dolichovespula* and *Vespula*. Thus, we concentrate on yellowjacket colony defense not only because most of the vespine literature concerns these wasps, but also because our research group (J. F. MacDonald, Purdue, A. Greene, University of Maryland, and P. J. Landolt, USDA, Miami, Florida) has been working on the yellowjackets of the Pacific Northwest (PNW) the past 10 years, providing us personal experience with their defensive behaviors.

This chapter deals only with vespine defense behaviors in relation to the colony and nest. It does not include data on individual wasps attacked by predators and parasites away from the nest, or information on the "manifestation of territorial defense" in which wasps of the same or different species will drive one another away from a feeding site (Free, 1970; Matsuura and Sakagami, 1973). Since the potential medical hazards of these venomous wasps have been discussed in recent reviews (Akre and Davis, 1978; Akre *et al.*, 1981; Davis, 1978; Ebeling, 1975), references to this aspect of vespine defensive behaviors are not included. Also, the behaviors during the stinging process are not discussed as they have not been fully investigated, although the mechanism of sting penetration was documented in Akre *et al.* (1981).

This chapter emphasizes vespine colony defense in relation to vertebrate and invertebrate nest intruders. Also covered are inter- and intraspecific behavioral interactions with other wasps, possible defenses against microbial agents, species differences in defensive behavior, and colony size and social organization in relation to nest defense. However, due to the limited published data on vespine defensive behaviors, this review is supplemented with personal observations obtained while excavating, collecting, transporting, and studying PNW yellowjackets.

BIOLOGY

All vespines build nests of masticated plant material, usually weathered wood, although soil is occasionally incorporated into some *Vespa* nests. The nests consist of several, horizontal combs attached below each other, which are typically covered with a multilayered paper envelope. Most species build nests either above the ground (aerial), or in subterranean sites, but a few species will utilize either location.

Although hornet populations in tropical areas are asynchronous, as colonies in various stages of development are present at any time of the year, all temperate *Vespa, Dolichovespula*, and *Vespula* have annual colonies. Colonies are haplometrotic (single foundress) and are usually monogynous (one queen). Inseminated queens overwinter in reproductive diapause and emerge during the late spring or early summer, depending on latitude and elevation. Queens feed on nectar and on captured prey, and their ovaries develop rapidly as diapause is broken. A queen soon selects a suitable nest site and gathers plant fibers to construct a nest. The queen or embryo nest ultimately consists of 20–65 cells usually covered by a paper envelope. During this stage the queen performs all the necessary colony functions, including oviposition and foraging for the developing brood. The queen at this time is extremely intolerant of conspecifics and other nest intruders. In about 30 days the first brood of workers emerges and takes over the duties of the colony, such as foraging for food and collecting construction materials for the nest. The queen rarely leaves the nest again, and her primary function is oviposition. Colony growth becomes exponential by midsummer. In late summer or early fall, depending on species, the colony constructs larger cells for reproductive production (new queens and males). Shortly after this time the colony enters decline, new males and queens emerge, and the colony eventually disintegrates. The males and queens mate, and the inseminated queens enter diapause in some protected location.

Since aggressiveness of a particular colony is often dictated by size, which in turn is a function of the number of worker-rearing combs, some general comparisons of nest size are warranted (Table 3.1). Hornet colonies are extremely variable in size and, since few data are available, generalities are difficult to make. However, most *Vespa* species probably have colonies smaller than species of the *Vespula vulgaris* group (*Paravespula* Blüthgen of some European workers, see Bequaert, 1931 for discussion of the species groups of *Vespula*). *Dolichovespula* colonies are usually small and are more comparable in size to those of the *Vespula rufa* species group. More detailed information on biology and taxonomy can be found in Akre *et al.*, (1982), Edwards (1980), and Yamane *et al.* (1980).

GENERALITIES

It is well known that among vespines only wasps inside the nest at the time of a disturbance participate in defense, and that returning foragers neither participate in defense nor communicate alarm to the nest inhabitants (Edwards, 1980; Gaul, 1953; Potter, 1964; Spradbery, 1973). The number of wasps responding to a disturbance is roughly proportional to the extent of the disturbance (Gaul, 1953), but each wasp apparently acts independently to locate the source of the disturbance (Gaul, 1953; Spradbery, 1973). A certain segment of the colony will not participate in defense until the envelope is torn away and the combs are exposed, and even then teneral individuals usually try to escape by crawling head first into empty cells. Occasionally even these callows will join in defense of the colony to bite and sting the intruder (Duncan, 1939).

The aggressiveness of individual colonies varies considerably with colony history. Colonies that have been disturbed in the past usually respond to disturbances more rapidly. Aggressiveness also varies with colony size, with small colonies being less pugnacious (Spradbery, 1973). Annual colonies of vespines are usually largest during reproductive production, and they are perhaps most alert and ready to defend the colony when these individuals are present. Defensive

TABLE 3.1. Colony Parameters of Vespine Wasps

		Mature colony size		
	No. of species	No. of workers	Total no. cells	No. workers' combs[a]
Vespa	22	Not available	600–14,000	Not available[a]
Dolichovespula	14	100–700	300–4,000	Most spp. single
Vespula rufa group	10	75–400	500–2,500	Single
Vespula vulgaris group	12	500–5,000	3,500–15,000[b]	Multiple

[a]Number of total combs (3–13) is quite variable between species. Single and multiple worker comb species are likely present.
[b]Not including rare, perennial, or polygynous colonies that may contain more than 120,000 cells.

flights of *V. pensylvanica* at this time will extend to 30 m or more.

Casual references to aggressiveness of vespine wasps usually also refer, directly or indirectly, to the size of the individual wasps, probably because large wasps are perceived to be more frightening than small ones. For example, workers of many *Vespa* are large, and the largest hornet, *Vespa mandarinia* Smith, is nearly always listed as the most aggressive species (Matsuura and Sakagami, 1973). However, no accurate technique has been developed to measure the intensity of colony defense of vespine wasps. Large colonies and large individuals are generally perceived to be the most belligerent. Consequently, casual references to levels of aggressiveness of various species (Horne, 1870; Pack-Beresford, 1902; Pagden, 1958; Rau, 1929; van der Vecht, 1957) or ranking of species as to aggressiveness (Chan, 1972; Matsuura and Sakagami, 1973) are quite subjective.

Aggressiveness of honey bee colonies has been quantitatively measured using a black leather ball as the object of attack (Michener, 1975). Since honey bees have large barbs on the sting and a preformed breaking point (Maschwitz and Kloft, 1971), stings of attacking individuals become lodged in the leather. Sting autotomy occurs frequently and the entire sting apparatus usually pulls free of the body (Hermann, 1971). Later, the number of stings embedded in the ball can be counted. Vespines have similar barbs on the lancets of the sting (Figure 3.1; Harwood and James, 1979; Edwards, 1980), but the sting rarely becomes embedded and pulled from the body. However, yellowjackets sometimes become lodged while stinging leather gloves, and although the sting is rarely pulled from the body, they remain entrapped. Perhaps some plastic or fabric could be used on a ball or similar object that would hook onto the barbs of the vespine sting. In addition, color (e.g., yellow) (Sharp and James, 1979) and movement of the ball might also be incorporated into the experimental methods. The number of stings or individuals on the test object could then be counted and serve as a reliable and comparative index of colony aggressiveness.

NESTING BIOLOGY IN RELATION TO DEFENSE

Nest sites undoubtedly provide the colony with shelter from some natural enemies. Aerial (arboreal) nests of *Vespa* and *Dolichovespula* are frequently constructed high in trees, which must afford them some protection from many vertebrate predators; however, these nests are cryptic and have been described as naturally concealed (Rau, 1929). A light brown or gray nest partially hidden from view by leaves

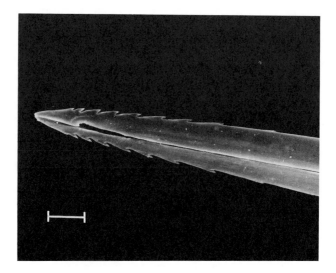

FIGURE 3.1. Sting lancets of *Vespula pensylvanica* worker. Line represents 0.1 mm. (Courtesy of A. R. Crooker, Electron Microscope Center,
Washington State University.)

and made more cryptic by shadows is nearly impossible to see. Nests
of other vespines are typically located in more protected locations in
the soil (subterranean), in tree stumps or cavities, or in wall voids of
man-made structures. Small colonies in these sites can be very
difficult to locate. Pack-Beresford (1902) was one of the first to
suggest that an ill-defined hole to the entrance tunnel of a subterranean nest might be a strategy to avoid predation. An extreme
example of cryptic entrances occurs in colonies of *Vespula acadica*,
where the hole rarely exceeds 1 cm (Reed and Akre, 1983a). Such
entrances are nearly impossible to detect when they occur in moss
on a decayed log in a dimly lit forest (Figure 3.2). No vespine is known
that actively camouflages its nest entrance, as reported for bumble
bees (Richards, K. W., 1978). However, many ground-nesting subterranean yellowjackets utilize abandoned rodent burrows (Mac-
Donald *et al.*, 1974; Akre *et al.*, 1982), and the nest is located in only
one of the many tunnels leading from the entrance hole. This
probably effectively "camouflages" the route to the nest and may
deter or at least confuse some predators. Nest sites in rodent burrows
also allow the queen several escape routes during severe colony
disturbances. For example, during the excavation of yellowjacket
colonies in the PNW, the queen sometimes leaves the nest by
retreating down one of the many interconnecting rodent burrows.

FIGURE 3.2. Cryptic nest entrance of *Vespula acadica* colony in a moss-covered log in northern Idaho.

The remaining workers usually rebuild the nest or construct a new nest within a few weeks. However, these strategies are primarily directed at visual-hunting predators and parasites. Most likely some vertebrate and most insect predators also utilize chemicals to locate prey.

Since it is possible that ground-nesting *Vespula* have chemicals marking the entrance (Butler *et al.*, 1969) or entrance tunnel, their predators may have the capabilities to utilize such pheromones as kairomones. Also, odors emanating from the colony or debris in the nest site may enable predators to locate the colony.

Although Yamane and Makino (1977) correctly state that vespine nests offer little defense against predators, the multiple sheets of envelope around a nest with a single entrance must be of some value in defense against invertebrate predators or parasites, or possibly invading microorganisms and fungi. Not only do vespines have a single aperature in the envelope, but ground nesters also have a single hole to their entrance tunnel that is constantly guarded by workers in large colonies (Figure 3.3). Such a system allows the colony to be easily defended.

Queen nests of *Vespa analis* Fab., *V. affinis* L., *Dolichovespula media* (Retzius), and *D. maculata* are frequently constructed with a long vestibule or funnel-shaped entrance spout (Figure 3.4; Yamane and

FIGURE 3.3. Mud turret around entrance hole of *Vespula pensylvanica* colony. Two guard workers are visible standing in the entrance facing outward.

Makino, 1977). While this could be an additional protection against predators and parasites (Yamane, 1976; Yamane and Makino, 1977), its role is probably more related to thermoregulation of the nest. The dead air space within the vestibule should serve as an excellent insulation. This tube is usually cut from the nest either by the queen just prior to emergence of the workers, or by the first workers themselves shortly after they emerge. A vestibule does not occur on the nests of more seasonally advanced colonies, but perhaps the air space between the lowermost comb and the bottom of the nest, sometimes consisting of over one-third of the nest, performs a similar thermoregulatory function (Greene *et al.*, 1976).

It has been suggested that the temperate vespine nest probably evolved to cope with environmental stresses (MacDonald, 1977), rather than ant predation as is probably the case for tropical vespid nests (Jeanne, 1975). While there are few recorded instances of ant predation on nests of *Dolichovespula* and *Vespula*, these omnipresent soil dwellers, nevertheless, must be of some importance in predation. For example, nests of colonies of *Dolichovespula arenaria* entering colony decline are sometimes robbed of all remaining brood by workers of *Formica obscurpies* Forel and other *Formica* species in Washington. Colonies of several ant species usually occur in the immediate vicinity of most underground vespine nests, and yellow-

FIGURE 3.4.   Queen nest of *Dolichovespula maculata* with a funnel-shaped entrance spout. The spout is almost severed from the nest.

jackets must be able to either physically exclude (i.e., nest envelope) these predators, or deter them by some unknown chemical means. Other social wasps such as species of *Mischocyttarus* (Jeanne, 1970), *Polistes* (Hermann and Dirks, 1974; Turillazzi and Ugolini, 1979), and possibly *Belonogaster* (Piccioli and Pardi, 1978) secrete a material on the pedicel of the embryo nest that is repellent to ants; however, no such secretion is known for vespines.

Another possible defensive mechanism is nest translocation. Two species of *Vespa* in Japan, *V. simillima* Smith and *V. crabro flavofasciata* Cameron, sometimes move from the original nest site to a new location (Edwards, 1980). The original nest is usually located in a small cavity, and when the colony becomes too large, the entire colony leaves to establish a new nest in a more spacious site. While this behavior could have possible defensive advantages (e.g., escaping brood parasites), it is possible that it concerns thermoregulation of the small nest. The small cavity has a microclimate more suitable for larval development in the embryo nest. Perhaps such absconding is a response to repeated nest disturbances or even partial destruction. Regardless of the function(s) this behavior serves, this phenomenon may explain the earlier reports of swarm founding by *Vespa* in tropical areas (van der Vecht, 1957). Nest translocation is unknown for *Vespula* or *Dolichovespula*.

Although the nesting associations of ants, wasps, and birds for defensive purposes against common predators are well known for polistine wasps in South America (Richards, O. W., 1978), nothing similar has been reported for vespines in either tropical or temperate regions.

BEHAVIORAL DEFENSE MECHANISMS

Defensive Postures

In one of the few reports mentioning vespine defense postures, Potter (1964) described the position of alerted *V. vulgaris* workers. Disturbed workers rise onto their tarsal tips with head held forward and abdomen arched, while constantly vibrating the wings. This position was similar to that of fanning workers, except that the wing beat was shorter and the frequency probably higher. This activity apparently alerted workers within the nest and caused them to run out the entrance tunnel. However, we have maintained colonies of *V. vulgaris* in glass-bottomed observation boxes (Akre *et al.*, 1976) for the last five seasons, but no posture could be deemed characteristic. Alert workers show variable behaviors, but many assume an elevated posture, their heads pointed at the disturbance, but they do not necessarily vibrate their wings. Workers of other studied species such as *V. pensylvanica, V. atropilosa* (Sladen), *V. consobrina* (Saussure), *V. acadica, Dolichovespula arenaria* (Fab.), and *D. maculata* also exhibit an alert stance or defensive posture that includes elevating the body; however, other characteristics of the position are too variable to make generalizations. Investigations into individual and colony (i.e, intra-specific) variation in vespine defense postures are needed before statements can be made concerning interspecific differences.

Defensive Flights

Defense flights, unless the source of the disturbance is readily apparent, are similar to orientation flights. The aroused wasps fly in circling paths that increase in ever-widening arcs as the defenders search for the intruder (Gaul, 1953). Detectable movement is rapidly investigated, and wasps concentrate their attack on fleeing nest invaders. Although Gaul (1953) stated that defensive flights were never more than 7 m from the nest, this distance varies among species and individual colonies. Rau (1929) gave a pursuing distance of 30 m for *D. maculata* (L.), the baldfaced "hornet." Workers from some colonies of *V. acadica* (Sladen), *V. pensylvanica* (Saussure), and *V. vulgaris* (L.) will pursue vertebrate predators at least 30 m from the nest. However, in most instances defensive flights are probably relatively short in both distance and duration. Apparently duration of these flights increases with a rise in temperature and may last from 90 seconds to 5 minutes or even longer on hot days (Gaul, 1953). Older reports such as that of Horne (1870), which stated that "*Vespa velutina*" will pursue a party for miles through the densest jungle, are undoubtedly in error. These reports probably arose from wasps becoming entangled in clothing and continuing to sting even though the nest disturbers had retreated from the nest site for considerable distances. Unknowingly disturbing additional colonies may also contribute to such exaggerated reports.

Defense of Embryo Nests

Defensive behavior of vespine queens on the embryo nest appears fairly typical for all studied species. When not foraging or engaged in other nest duties, the queen is usually curled about the nest pedicel, resting, but perhaps also incubating the brood (Brian and Brian, 1948). When the nest is slightly vibrated, the queen rapidly rushes out and quickly walks about the envelope (Yamane and Makino, 1977). Queens of *Vespa analis insularis* Dalla Torre (Yamane and Makino, 1977) and of *Dolichovespula maculata, D. arenaria, Vespula acadica, V. atropilosa,* and *V. pensylvanica* also exhibit this patrolling behavior. Also, yellowjacket social parasites and occasionally some queens (e.g., *V. acadica*) perform a flailing behavior while perched on the comb or envelope (see Figure 5 in Greene *et al.*, 1978) which appears to be a defensive action or threatening response to an intruder. During a more violent disturbance the queen rapidly flies from the nest and circles the intruder. In the case of vertebrate intruders, particularly humans, the queen rarely stings or even attempts to sting during such disturbances. Instead, after circling the intruder the queen usually leaves the nest site and is gone 15 minutes to several hours. Yamane and Makino (1977) reported the unusual behavior of alerted *Dolichovespula* queens leaving the nest and suddenly dropping to the ground. No further information is available on this behavior, and it has not been observed for queens of PNW yellow-jackets.

Sanitation Behavior with Respect to Defense

Yellowjackets of the *Vespula rufa* species group have untidy nests with envelope fragments of prey remains, dead inhabitants, and drops of excrement scattered about the nest cavity and on the nest itself (Akre *et al.*, 1976, 1982). These studies also indicated that queens as well as the workers of *V. pensylvanica* (*V. vulgaris* group) were more fastidious in sanitation tasks than queens or workers of *V. atropilosa* (*V. rufa* group). This behavior also seems to correspond to their response to invertebrate nest intruders. Arthropods entering a nest of *V. pensylvanica* are immediately and relentlessly attacked until they are killed or driven from the nest. In contrast, intruders into the nest vicinity of *V. atropilosa* (also *V. acadica*) are sometimes vigorously attacked by some members of the colony, but they are just as often ignored, or only briefly attacked. This intermittent or partial nest defense by only a few workers might also extend to defense against vertebrate intruders. Previously undisturbed colonies of *V. rufa* group species are generally considered not as easily aroused, or docile in comparison to other yellowjackets.

The poor sanitation by *V. rufa* group species can perhaps be considered a method of nest defense. This dictates that the workers make fewer trips in and out of the cryptic entrance, which possibly lowers the chances that visual-hunting enemies can find the colony. There are no quantitative data available to support this contention, but few colonies of *V. rufa* group species have been reported in the literature (see Akre *et al.*, 1982), probably because of the difficulty in locating colonies.

Dolichovespula nests are comparable to *V. vulgaris* group species with regard to sanitation, and at least *D. arenaria* (Greene *et al.*, 1976) and *D. maculata* workers are readily aroused to attack nest intruders. No information is available on sanitation in *Vespa* colonies, but from the lack of references to untidy nests, it is assumed their sanitation behavior must be comparable to *Dolichovespula* and the *V. vulgaris* group. However, many *Vespa* have not been studied, and it is probable their sanitation and defensive behavior varies greatly among the 22 species and numerous subspecies.

Guard Workers

Certain workers are more active in colony defense than others. These guard workers quickly become excited by vibrations around the nest or entrance, and frequently attack and sting the intruder(s).

A study of colonies of *V. atropilosa* and *V. pensylvanica* in glass-bottomed observation boxes showed that in incipient colonies of less than 10 workers, guards comprised only one or two workers standing on the envelope while facing the entrance (Akre *et al.*, 1976). As more workers emerged, one or two positioned themselves adjacent to the nestbox entrance and antennated returning workers. Eventually more workers served as guards, until the entrance was surrounded, by as many as 20 workers in mature *V. pensylvanica* colonies (Figure 3.3). Apparently, the number of guard workers is roughly proportional to colony size. For example, *V. pensylvanica* colonies are more populous than *V. atropilosa*, and subsequently have more guards. *V. consobrina* (Saussure), another *V. rufa* group species along with *V. atropilosa*, averaged 2.5 guards in the entrance at any one time (Akre *et al.* 1982). The number of guards varied by time of day with the greatest number present from 8 to 10 PM. Colonies of *V. vulgaris* have more guard workers in the morning before foraging begins, and in the late evening when foraging wanes (Potter, 1964). Numerous guards are also present during inclement weather, when few workers are able to forage.

Potter (1964) found that older workers, especially those greater than 30 days old, tended to serve as guards. Older workers in colonies of *V. pensylvanica* and *V. atropilosa* are not necessarily guards, but perform few colony tasks (Akre *et al.*, 1976). They do not forage, engage in little nest construction, and rarely feed larvae. However, some workers apparently specialize in certain tasks, and some are primarily guards throughout their lives (Akre *et al.*, 1976, 1982).

Colonies of *V. rufa* group species of yellowjackets (e.g., *V. acadica, V. atropilosa*, and *V. consobrina*) lack guard workers at the exterior entrance to their nest, even late in the season, when the mature colony contains 100 or more workers. Guards are frequently at the end of the entrance tunnel where it opens into the nest cavity. However, colonies of *V. pensylvanica* or *V. vulgaris* have guard workers completely surrounding the entrance hole from the end of July onward (Figure 3.3). The number of guards reaches a peak at 8–10 PM, after which most workers retreat further down the tunnel, perhaps to the nest. In the early morning, prior to initial foraging activities (4–6 AM), guards return to the entrance. Late season colonies of *V. pensylvanica, V. vulgaris*, and other *V. vulgaris* group species consist of 1,500 to 4,000 or more workers, and a disturbance in the vicinity of the nest causes an immediate eruption of workers from the entrance tunnel.

The aerial nesting yellowjackets, *D. arenaria* and *D. maculata* have guard workers that remain inside the nest but adjacent to the entrance hole (Figure 3.5). The slightest nest vibration causes them to either

FIGURE 3.5. Two guard workers of *Dolichovespula arenaria* standing just inside the nest entrance.

fly out to attack, or walk quickly out on the envelope in an alert defensive position. Since colonies of these species are relatively small (Greene *et al.*, 1976; Akre *et al.*, 1981), guards are not numerous, probably no more than four to six.

Although little information is available on guarding behavior in *Vespa* colonies, this activity appears to be quite similar to the pattern in yellowjackets. The number of guards in *V. orientalis* colonies increases with size, and like colonies of the *V. vulgaris* species group, large colonies may have up to 20 guards (Ishay *et al.*, 1976). A *V. orientalis* guard may remain at the entrance for 7–12 minutes. A replacement appears nearly immediately after the guard returns to the comb. The new guard circles outside the nest entrance and then remains posted just outside the entrance. Also as in yellowjackets, guards may be present at night, usually inside the nest, but in one species, *Vespa mongolica* André (*Vespa simillima* Smith), workers remained on the envelope throughout the night (Matsuura, 1973).

Species Differences

Few differences have been observed in defensive behavior among yellowjacket species. Most species exhibit similar defense behaviors with most differences attributable to colony size or stage of colony growth. However, some species perform apparently unique positions or actions during colony disturbances. For example, workers of *V. acadica*, when alerted by nest vibrations, become immobile on the comb or envelope (Reed and Akre, 1983). If the disturbance is intermittent, the entire outer envelope of the nest is eventually occupied by immobile workers facing outward and raised on their tarsi. The antennae are still and outstretched. This immobility does not last long until the workers resume activity or begin to investigate the causes of the disturbance. This "frozen" stance has not been seen in colonies of other yellowjackets. Other species appear to actively run about the nestbox during disturbances.

Gaul (1953) reported an unusual defensive behavior for colonies of *V. consobrina*. When a colony was disturbed, workers flew less than 10 feet away, and then dropped to the ground. Then they crawled up onto the intruder to sting. However, this behavior was not encountered in collecting colonies in the PNW (Akre *et al.*, 1982).

There are also subtle differences in defensive behavior between workers of *D. arenaria* and *D. maculata*. A disturbed nest of *D. arenaria* results in aroused workers flying out of the entrance hole to attack, but apparently few crawl out on the envelope. However, in some *D. maculata* colonies, many workers crawl out on the envelope with their bodies elevated and pointed toward the disturbance (Figure 3.6). Relatively few workers fly straight out to attack. A similar response to disturbances occurs in *Vespa affinis* (L.), as the workers rapidly patrol the envelope but do not readily fly out and attack. This response is apparently different from that in more aggressive *Vespa* species (Matsuura, 1973). In addition, *D. maculata* workers will circle a human intruder, and sometimes hover, motionless in the air with gaster curled forward, before darting in to sting. This ability to hover has been reported for *V. acadica* (Reed and Akre, 1983) and *V. consobrina* (Akre *et al.*, 1982) while entering the nest entrance or during foraging. Also, this hovering was quite noticeable during nest disturbance and subsequent defense in some *V. acadica* colonies.

Dolichovespula also seem to differ in their attack behavior on vertebrates. While most yellowjackets attack at any part of the intruder, *Dolichovespula* concentrate their attacks on the head and upper body (Greene *et al.*, 1976). This point of attack may be related to their normal foraging behavior, since *Dolichovespula* often forage around large animals, perhaps searching for flies, and tend to persistently circle around the animal's head (Greene *et al.*, 1976). *Vespula* rarely show this foraging behavior, and when performed, they only make one or two circles and then leave.

FIGURE 3.6. Workers of *Dolichovespula maculata* responding to a disturbance by rushing out of the nest envelope.

AUDITORY ALARM/DEFENSE

Most reports of vespine nest defense indicate that the inhabitants become alarmed by vibration of the nest or by the entry of intruders (Edwards, 1980; Gaul, 1953). The response shown by nest inhabitants may be stimulated by sounds produced by wasps vibrating their wings in the defensive posture (Potter, 1964). This sound, coupled with a visual stimulation, may be responsible for most defensive reactions. Auditory cues for eliciting nest defense are quite plausible in view of vespine structure, which readily transmits solid-borne sounds and facilitates transmission of airborne sounds (Ishay and Schwartz, 1973), and because acoustic signals are commonly used in communication within the colony (e.g., larval "scraping," abdominal drumming, Ishay, 1977). Other social insects also use sound in colony defense. For example, guard workers of honey bee colonies (*Apis cerana* Fab., *Apis mellifera* L.) produce a sound via wing vibration when attacked by invading hornets (Ohtani and Kamada, 1980; Matsuura and Sakagami, 1973), which possibly alerts more colony defenders. However, it has also been reported that vespines such as *Vespa orientalis* Fab.

(Ishay *et al.*, 1967, *Vespula germanica* (Fab.) and *V. vulgaris* (Maschwitz 1964a, b) have alarm pheromones.

Workers of *Vespa mandarinia* often fly around nest intruders while making a loud clicking sound before attacking (Matsuura and Sakagami, 1973). This sound is produced by rapid mandibular movements. It is apparently produced by all Japanese hornets in or near their nests and is a very prevalent defense mechanism of other *Vespa* wasps. Unfortunately, it is not known whether this act communicates alarm to other colony members or if it is an effective deterrent to intruders.

CHEMICAL DEFENSE

Alarm Pheromones

Most literature on nest defense by vespines (i.e., Gaul, 1953; Potter, 1964; Spradbery, 1973) indicates that the wasps do not communicate alarm to their nestmates, or if they do, it is probably by means of sound produced by workers giving the defense reaction. However, Maschwitz (1964a, b) presented evidence for an alarm pheromone in *Vespula germanica* and *V. vulgaris*. A worker was seized and held in the nest entrance, causing a number of wasps to become alarmed and rush out. A similar effect was obtained by using only a squashed sting. The pheromone component was reported as smelling like fermented fruit. Similar experiments using crushed workers and acetone extracts of stings failed to release alarm behavior in colonies of *V. atropilosa, V. pensylvanica,* and *V. vulgaris.* Conversely, the slightest vibration of the soil around the nest produced an immediate defense response with workers rapidly issuing from the entrance hole to attack. More recent experiments with *V. vulgaris* colonies in England have indicated that these wasps have an alarm pheromone associated with the sting (Edwards, 1980). This material, presumably released only inside the nest, causes the wasps that detect it to fly outside to attack. Although various alkanes and ketones elicit alarm behavior in *Vespa orientalis*, the Oriental hornet (Saslavasky *et al.*, 1973), these researchers did not demonstrate that these chemicals were indeed produced by excited workers. Also, according to these researchers, hornet workers only spray venom, presumably containing alarming substances, outside the nest.

Venoms

The most effective vespine defense is the ability to sting repeatedly and inject venom into a potential intruder. Vespids possess a venom ejection mechanism termed the "spray sting type" by Maschwitz and Kloft (1971). Venom is ejected from the poison gland reservoir by muscular contraction, rather than by pumping values in the sting lancets as in most aculeate Hymenoptera. Therefore, in addition to venom injection during sting penetration, it is at least theoretically possible that all vespids can spray venom. However, only workers of *D. arenaria* (Greene *et al.*, 1976), and *Vespa orientalis* (Saslavasky *et al.*, 1973) have been reported as spraying venom against nest intruders. There are no reports of venom sprayed at invertebrates and this probably would be an ineffective deterrent. However, venom sprayed into the eyes of vertebrates, at least humans, causes intense urtication and tears. Thus the intruder is rapidly, albeit temporarily, blinded and distracted.

Schmidt and Blum (1979) investigated the toxicity of *D. maculata* worker venom to mice, conspecifics, and other insects. They concluded that the venom was not highly toxic to any of the insect species tested, unless injected directly toward a ganglion, and that the primary role of the venom was for use against vertebrate predators. This research revealed that teneral wasps produced essentially no venom. This explains, at least in part, the escape reaction of callows when the nest is invaded. Evidence was also presented indicating that the venom contains a short-term, nonlethal toxin that rapidly paralyzed conspecifics. Within 10 minutes or so, the wasps recovered only to subsequently become paralyzed again and die. A similar reaction was observed in a *D. maculata* queen repeatedly stung by workers. The queen became paralyzed, recovered in less than 10 minutes but was found dead the following day.

Venoms of the social parasites *Dolichovespula arctica* (Rohwer) and *Vespula austriaca* (Panzer) have not been chemically investigated, but observations of behavioral interactions between these parasites and their host species, *D. arenaria* and *V. acadica*, respectively, indicate the venoms are very toxic to the hosts (Greene *et al.*, 1978; Reed, 1981; see also Schmidt *et al.*, 1984). Workers stung by the parasite are instantly paralyzed, fall from the nest, and never move again except for a spastic twitching of the tarsi. The reaction of host queens (e.g., *V. acadica*) to a sting is nearly as sudden, as they also quickly become paralyzed and die. Evidently, the venom of the facultative social parasite *V. squamosa* also causes an immediate death of the host queen, *V. maculifrons* (Matthews and Matthews, 1979). Also, a *V. austriaca* female was stung, possibly more than once by *V. atropilosa* workers, and then quickly fell to the glass paralyzed, but recovered within 90 seconds. The parasite was found dead the next day, but it cannot be ruled out that workers stung her again when the colony was not being observed. If the parasite was not stung again, this delayed reaction to the venom of workers is very similar to that reported for *D. maculata* venom (Schmidt and Blum, 1979).

The sting is a defensive weapon only and it is rarely or never used in prey capture (Duncan, 1939; Gaul, 1941; Schmidt and Blum, 1979; Akre, 1982). However, workers of several species of PNW yellowjackets will attempt to sting if they attack an arthropod that physically overpowers or injures them during the attack.

VERTEBRATE DEFENSE

Yellowjacket colonies are attacked and consumed by bears (Bigelow, 1922; Buckell and Spencer, 1950; MacDonald and Matthews, 1981) (*Dolichovespula arenaria, Vespula consobrina, V. pensylvanica*), skunks (Preiss, 1967) (*V. maculifrons,* Buysson), raccoons (MacDonald, 1977) *V. maculifrons,* coyotes (Akre *et al.,* 1981; 1982) (one colony of *V. consobrina* and *V. vulgaris*), badgers, (Spradbery, 1973; Edwards, 1980), and birds (Birkhead, 1974; Gabriel and Gilbert, 1975; Cobb, 1979). Although there is no quantitative information, some small colonies are probably destroyed by small rodents such as mice and ground squirrels (Spradbery, 1973; Roush and Akre, 1978; Edwards, 1980).

Bears attacking colonies of yellowjackets evidently suffer some pain, as they often stop to snarl and roll on the ground, but immediately reattack the colony (Bigelow, 1922). There are no observations of reactions of raccoons, skunks, or badgers to stings, or mention of any special means of colony defense (i.e., concentrate attack on vulnerable sections such as eyes or nose) against vertebrate predators.

Several species of birds will catch and use worker wasps as food (Gabriel and Gilbert, 1975; Birkhead, 1974). However, Birkhead (1974) suggested that the small number of bird species that have been reported as attacking wasp nests indicates that nest defense is highly effective. In addition, only a few species have been reported to regularly attack nests. One of these, the honey buzzard, is evidently persistent in its attack. Cobb (1979) gave a detailed account of a honey buzzard excavating and attacking a subterranean *Vespula* colony in England. Although driven from the nest numerous times by the defending wasps, the bird managed to reach the nest and consume the brood combs over a period of 6 days (see also Edwards, 1980).

Although detailed records have not been kept of the number of subterranean *Vespula* colonies destroyed by animals in the Pacific Northwest, it appears predation is light. Perhaps one to four of 15–50 located colonies are destroyed each year. In eastern Washington most colonies, primarily *V. atropilosa*, were destroyed by striped skunks. No mature *V. pensylvanica* colonies, which usually contain 1,000 or more workers, have been attacked while still active. However, skunks will readily dig up and consume large colonies that have been killed for subsequent excavation and analysis. In the forests of northern Idaho and northeastern Oregon an occasional colony of *V. acadica* or *V. vulgaris* was found that probably was attacked and partially eaten by a bear or coyote. No attacks were witnessed, but tracks and scats of these predators were present in the immediate vicinity.

Predation in other areas of the United States may be heavier. Thirteen of 27 *V. maculifrons* colonies in Athens, Georgia, were destroyed by raccoons (MacDonald, 1977). Additionally, during a 2-year study in Delaware nearly 80% of the *V. maculifrons* colonies (119 of 151) were destroyed by skunks; another two colonies were destroyed by moles (Preiss, 1967).

Records of nest defense by *Vespa* against mammalian predators are not available. Perhaps this is simply because it rarely occurs. The large worker size, in combination with populous mature colonies, probably deters most potential predators. Besides, many hornet colonies are well known for their aggressiveness (van der Vecht, 1975; Chan, 1972).

INVERTEBRATE DEFENSE

Hornet Predation

The giant hornet, *Vespa mandarinia*, not only attacks and occupies honeybee hives, but also preys upon *Vespula* and *Vespa* colonies (Matsuura and Sakagami, 1973). As few as ten hornet workers can successfully invade a colony of *Vespula lewisii* (Cameron) within 1 hour with the loss of only one or two hornets. Defending yellowjackets battle the attackers, as evidenced by dead hornets at the nest entrance, but their counterattack apparently is not as intense as that of *Apis mellifera*. The giant hornet also attacks colonies of other hornet species such as *V. simillima, V. crabro* L., and *V. analis* Fab. The successful attack usually undergoes three distinct phases similar to that of predation on honeybee hives. The attack begins with a hunting phase, when a giant hornet captures an individual hornet worker near the nest entrance. The attacker usually kills the victim with its mandibles and malaxates the hornet into a ball for transport back to the colony. However, the prey hornet sometimes kills the attacking *V. mandarinia* by using its sting. Large colonies of *Vespa crabro* and *V. simillima* resist hunting giant hornets when several defending workers fly out and chase the invader(s) away. In these cases, the hunting phase is interrupted and never proceeds to the following slaughter phase where several attackers systematically kill defending workers but do not utilize them as prey items. During this phase, combatants fall to the ground locked in violent battle using legs, mandibles, and stings. This phase may last several hours or until the following day. After a successful slaughter the giant hornets occupy the prey colony and carry the brood back to their colony. However, small colonies usually do not resist the attackers but may abandon the nest, and thus serve as easy prey for giant hornets.

Other hornets such as *Vespa crabro* (DeJong, 1979) and *V. orientalis*, an important predator of honeybee colonies in the Middle East (DeJong, 1978), probably occasionally raid yellowjacket colonies. In contrast, there are no substantiated reports of any *Vespula* preying on an active hornet colony, although Ishay *et al.* (1976) stated that weakened colonies of *V. orientalis* are sometimes invaded by *V. germanica. Dolichovespula maculata* workers attack *Vespula* individuals in the field (Howell, 1973), but it is not known whether the bald-faced hornet is capable of invading active *Vespula* colonies.

Intra- and Interspecific Competition

Intra- and interspecific competition among yellowjacket and hornet queens for nests is probably a frequent occurrence, with intraspecific competition being more common (Akre *et al.*, 1977; Matthews and Matthews, 1979). Akre *et al.* (1976), MacDonald and Matthews (1975), and Matsuura (1970) reported that dead queens were often found in the entrance tunnels of subterranean nests. Many queens are observed seeking nesting sites after other conspecifics have already established nests. However, a foundress queen is highly intolerant of other queens and vigorously attacks any invading queen (Akre *et al.*, 1976; Matthews and Matthews, 1979). The resident queen wins most confrontations (Matthews and Matthews, 1979) and thus, probably most of these invading queens are killed while trying to usurp nests.

Direct and indirect evidence for intense intra- and interspecific competition among *Vespula vulgaris* and *V. pensylvanica* queens was found during the summer of 1980 in Field's Spring State Park in southeastern Washington. In the entrance tunnel of several colonies of *V. vulgaris*, there were remnants of conspecific queens as well as *V. pensylvanica* queens, which periodically usurp *V. vulgaris* colonies (Akre *et al.*, 1977; Roush and Akre, 1978). More direct evidence was provided by an excavated colony that not only had four dead *V. vulgaris* queens in the entrance tunnel, but also contained two viable *V. vulgaris* queens (Akre and Reed, 1981a). It is possible that the nest usurpation process was still underway or that there was a brief time of peaceful coexistence. Another *V. vulgaris* colony was detected on July 11 while a possible usurpation attempt was in progress. When discovered, three guard workers were in front of the nest entrance very alert and agitated. Soon a *V. vulgaris* queen flew in and perched on a dead branch hanging 3–4 cm above the nest entrance. While remaining there, she intermittently buzzed her wings about every 4–6 seconds. During the next 5 minutes she attacked seven workers as they flew from the entrance, knocking them to the ground. Sometimes both the queen and worker fell to the ground locked in struggle. However, they always quickly broke apart and the queen returned to her "perch." During the next 30 minutes the queen remained on the branch while buzzing her wings, and attacked two more existing workers. Due to the rapid actions we were not able to determine whether or not the queen or worker attempted to sting, but none of the workers seemed harmed. Although this queen was marked, she was not recaptured at the entrance or inside the colony.

Other researchers have also discovered that competition for nests among queens is relatively frequent. Matthews and Matthews (1979) determined from marked foundress queens [mostly *V. maculifrons* (Buysson)] of embryo nests that usurpation rates ranged from 23% to 37% annually over a 3-year period. Multiple takeovers were common and most attempts involved conspecifics.

In some instances nest construction patterns provide reliable evidence of interspecific nest usurpation. A transition in fiber type and color occurs in nests of *V. vulgaris* usurped by a *V. pensylvanica* queen (Akre *et al.*, 1977; Roush and Akre, 1978). After the *V. pensylvanica* workers emerge, they apply grey fiber to the existing *V. vulgaris* nest and thus produce an abrupt change from the tan, brittle combs or envelope of the original nest to the more pliable, grey paper of *V. pensylvanica*.

One such usurped or adopted *V. vulgaris* colony provides an example of the interactions between the two species. The nest consisted of three combs with a total of 624 cells and contained about 60 workers (44 were collected). A queen of *V. vulgaris* and *V. pensylvanica* were present together in the nest when collected. The colony was transplanted into a glass-bottomed nest box (Akre *et al.*, 1976) for observing colony behavior. The next day, the *V. pensylvanica* queen was first noticed while the *V. vulgaris* queen was dead on the glass with her head severed. During July 11–23 the *V. vulgaris* worker force decreased from 44 to 4, while the *V. pensylvanica* workers increased from 0 to 204. When workers of both colonies were present, interspecific worker trophallaxis commonly occurred. In most cases, a *V. vulgaris* worker was the donor to a *V. pensylvanica* worker. Fiber loads were brought in by workers of both species (tan by *V. vulgaris*, grey by *V. pensylvanica*) and were placed side by side on the envelopes or combs. Workers of both species returned with prey loads and shared them with workers of the other species. The only possible sign of conflict between the workers was the high frequency of maulings of the *V. vulgaris* workers by those of *V. pensylvanica* (a stereotyped interaction in which one worker bites another with no apparent harm inflicted on the usually motionless recipient; see Akre *et al.*, 1976). In these instances the mauling was particularly vigorous and long (greater than 3 minutes), and the maulee tried to crawl away instead of typically curling up and remaining motionless. No sting attempts by maulers were clearly observed and no maulee was visibly injured.

Excluding social parasites (see following section) the only other records of interspecific competition involve nests of *V. vulgaris* or *V. maculifrons* usurped by queens of *V. flavopilosa* (MacDonald *et al.*, 1980). However, *V. germanica* may occasionally take over *V. vulgaris* colonies in Europe, as Nixon (1935) found one *V. vulgaris* nest usurped or adopted by a *V. germanica* queen. Interspecific usurpations by non-parasitic queens have not been reported in species of *Vespa, Dolichovespula*, or the *V. rufa* group. However, this may be only because of the lack of knowledge of the nesting biology of these vespines.

Nest usurpation usually occurs early in the nesting season when a lone queen, or a queen accompanied by only a few workers, is present to defend the colony. The resident queen uses her sting as well as mandibles in confronting any potential usurper. The workers also attack and mob any foreign queen entering the nest using their mandibles and stings. The mandibles may be the more important defensive weapon of the workers as they may hold down the larger intruder by grasping its appendages and thus impede the usurper's progress (see figures in Matthews and Matthews, 1979; and Figure 73 in Edwards, 1980). This defensive strategy may also allow other workers to gain better purchases to sting the intruding queen. Similar defensive responses are observed when social parasites invade a colony. Undoubtedly due to the intense defensive reaction of the workers, usurpation must become increasingly difficult as the worker force increases.

The sting of the queen appears to be reserved for combat with other intruding queens as she rarely attacks or stings during other nest disturbances. The toxicity of queen venom has not been throughly investigated; however, Schmidt et al. (1980) found that the venom of V. maculifrons queens was only one-sixth as active against mice as worker venom. They suggested that this reflected the role of the worker caste in colony defense against vertebrate predators. Since queens are frequently involved in mortal combat with other yellow-jackets trying to usurp their nests, toxicity of queen venom may differ from worker venom for invertebrates and perhaps is especially toxic to other queens.

Colony Drift

Although workers readily attack or repel any foreign yellow-jackets attempting to enter the colony, some conspecific workers, or those of a different species, occasionally may join other colonies (Edwards, 1980). For example, one or two V. consobrina workers joined a colony of V. atropilosa and reproduced some males (Akre et al., 1976). Similarly, V. atropilosa workers sometimes joined another V. atropilosa colony with little or no fighting (Akre et al., 1977), although in some cases intruding workers are routinely mauled by the workers. Balduf (1968) reported on two colonies of V. vulgaris that contained V. maculifrons workers, but these might have been colonies usurped by a V. maculifrons or V. flavopilosa queen. However, conspecific colony drift evidently occurs in V. maculifrons (Lord et al., 1977). Conversely, a worker of V. pensylvanica that entered a V. consobrina colony in a glass-bottomed observation box was repeatedly attacked. It was success-fully driven away only after 2 days. This suggests that workers may more readily enter colonies of conspecifics or closely related species (e.g., V. consobrina and V. atropilosa belong to the same species group) rather than colonies of more distantly related species.

Social Parasitism

Facultative, temporary social parasites include vespines that regularly usurp colonies of other species and rear their own workers, which eventually supplant the decreasing force of the host species (Taylor, 1939). However, these species still retain the ability to initiate their own colony without utilizing a host nest. In North America, *Vespula squamosa* (Drury) is parasitic in this fashion on *V. maculifrons* (MacDonald and Matthews, 1975; 1981), *V. vidua* (Taylor, 1939), and *V. flavopilosa* (MacDonald *et al.*, 1980). The existence of transition zones in nest construction was one of the initial evidences to alert MacDonald and Matthews (1975) to the usurpation of nests of *V. maculifrons* (tan, brittle paper) by queens of *V. squamosa* (grey, flexible paper). Subsequent investigations in Georgia during 1974 and 1975 (Akre *et al.*, 1981) showed that 99 of 114 *V. squamosa* nests had originally been initiated by *V. maculifrons* but had later been usurped by this social parasite. Apparently, *V. squamosa* invasions peak when the host nest averages 278 cells and probably contains about 20–30 workers (Matthews and Matthews, 1979). However, some usurpers probably invade earlier, when they would have to contend with fewer workers. An introduced *V. squamosa* queen is attacked by the workers, which grasp appendages in their mandibles and attempt to sting (Matthews and Matthews, 1979). The usurper is about three times the size of the *V. maculifrons* workers and thus is able to bite them in two through the abdomen or cervical areas. The alien and resident queen also fiercely battle each other with the sting serving as the principal weapon. Thus, defense by the host colony members appears to be similar to colony responses toward any alien queen.

Another facultative, temporary social parasite is the hornet, *Vespa dybowskii* Andre. This species sometimes constructs its own nest, but also usurps the nests of *Vespa crabro* and *V. xanthoptera* Cameron (Sakagami and Fukushima, 1957). Females of this species are particularly aggressive and have stronger sclerites than other *Vespa* species. The foundress of the host species is killed at the time of the invasion or shortly thereafter. The parasite has a strong odor, but it is unknown if this plays any part in usurpations. There are no published accounts of nest defense by the foundress.

The categories of obligate, permanent, and social parasites include yellowjackets and hornets that never construct their own nests but instead usurp those of a closely related species early in the season (Taylor, 1939). These parasites lack a worker caste and use the host workers to rear their reproductive brood. Although obligate vespine social parasites occur in Europe and Asia (Akre, 1982), no behavioral observations of nest defense have been recorded. However, studies of *Dolichovespula arctica* (Rohwer) and *Vespula austriaca* (Panzer) in North America include information concerning nest usurpation interactions between these parasites and their hosts. *Dolichovespula arctica* is found in the nests of *D. arenaria* and *D. norvegicoides* (Sladen) (Greene *et al.*, 1978; Jeanne, 1977; Yamane, 1975), whereas *V. austriaca* has been found in the nests of *V. acadica* in North America (Reed *et al.*, 1979).

Although it might be assumed that this high-risk reproductive strategy rarely occurs, Jeanne (1977) found 13 of 20 (65%) *D. arenaria* colonies in central New England, 1972–1976, infested with *D. arctica*. Five of seven (71%) postemergence (workers present) and eight (62%) pre-emergence colonies were parasitized. It was estimated that the actual rate of parasitism in the population was closer to 71%. Collections of 106 *D. arenaria* colonies in areas of the Pacific Northwest (i.e., southeastern Washington, northeastern Oregon, northern Idaho) during 1976–1980 showed an infestation rate of about 22% ($N = 106$). Nearly 19% of the queen nests were infested ($N = 16$), while 22% of the postemergence colonies were parasitized ($N = 90$). Although not as high as Jeanne's data, these rates are still high enough to regard this social parasite as a primary "predator" on colonies of *D. arenaria*. *Dolichovespula arctica* probably has a more serious impact on the population of the host species than vertebrate predators.

A *D. arctica* female enters a queen nest or a nest with a few workers and is usually attacked by the colony member(s) (Figure 3.7) (Evans, 1975; Greene *et al.*, 1978; Jeanne, 1977). Surprisingly, the parasite seldom fights back, but instead becomes passive (Greene *et al.*, 1978). The inquiline is presumably protected from stings by thick sclerites and close-fitting segments common to most vespid social parasites (Greene *et al.*, 1978; Wilson, 1971). However, if the attack becomes too vigorous, the parasite will leave the nest, but may soon return. This process of colony attack, and subsequent departure, may continue repeatedly. The attacks gradually subside and the parasite is accepted into the colony. From this point onward, the *D. arctica* is dominant in all interactions with colony members (Greene *et al.*, 1978). The host queen and inquiline initially coexist, but later, at least in some colonies, the queen is killed by the parasite. Although workers show some attraction to the *D. arctica* and frequently nibble its body, interactions become more aggressive. Occasionally the parasite kills many of the workers (Greene *et al.*, 1976) or, in contrast, is killed by the workers (Evans, 1975) or driven from the nest.

FIGURE 3.7. A female social parasite, *Dolichovespula arctica*, and a worker of the host species, *D. arenaria*, on the nest envelope.

The rapid pacification of the foundress and workers during initial colony invasion suggests that a chemical may mediate this behavior. The parasite drags its gaster on the comb as it walks, suggesting that this material may possibly be produced by sternal glands of the abdomen (Greene *et al.*, 1978; Jeanne, 1977). *Dolichovespula arctica* females also have Dufour's glands four to eight times the length of the same gland in host queen (Jeanne, 1977), and this may be the source of a pacifying chemical. The parasite also grooms her abdomen very frequently during her initial occupation period (Jeanne, 1977). This may enable the inquiline to acquire the colony odor or to aid in the dispersion of a pacifying chemical from exocrine glands. Although it has been stated that the facultative social parasite *Vespa dybowskii* has a peculiar odor (Sakagami and Fukushima, 1957), no mention was made of this playing a role in the nest invasion process. An additional suggestion that pacifying chemicals might be used by social parasites can be inferred from the studies of Francke *et al.* (1978). They isolated volatile spiroketals from the abdomen of *V. vulgaris* workers and proposed that they might serve as aggression inhibitors protecting a wasp from attack by its nest mates. Perhaps these inquilines use such chemicals in the invasion process.

The other social parasite occurring in North America, *V. austriaca*, has been reported from nests of *V. acadica* in North America, but also usurps colonies of *V. rufa* in Europe and Asia (Reed *et al.*, 1979). Collections of *V. acadica* colonies in northern Idaho and in the Blue Mountains of southeastern Washington during 1978–1980 showed 7 of 19 (37%) nests were parasitized (Reed, 1982; Reed and Akre, 1983a). Parasite populations are probably this high only in localized areas, since previously collected colonies of *V. acadica* were not parasitized (Roush and Akre, 1978). Unlike *D. arctica*, which usually invades a host nest containing only the queen, *V. austriaca* enters the nest of its host when workers are present (Reed and Akre, 1980; Reed, 1982; Reed and Akre, 1983c). Also, there is apparently no period of peaceful coexistence, but the parasite female, host queen, and workers engage in a violent combat. The parasite does not always win, but may be driven from the nest or perhaps killed. However, if the host queen is killed or chased away, the parasite female is dominant in all interactions with the remaining colony members.

It is possible that the cryptic nest entrance of this host species may serve as a "defense" against *V. austriaca*. However, it is equally plausible that social parasites find the nest by using the chemical marking the entrance (Butler *et al.*, 1969; Matsuura, 1970) or a pheromonal trail leading to the nest. The bumblebee social parasites, *Psithyrus* spp., evidently locate subterranean host colonies by odor (Cederberg, 1979) and thus it appears likely that vespine social parasites have a similar host-finding mechanism.

Other Nest Associates

Although vespine colonies provide a suitable habitat for many arthropods (MacDonald *et al.*, 1974, 1975; Spradbery, 1973), most of these nest associates are encountered primarily during colony decline. Most are scavengers feeding on debris below or inside the nest and do not adversely affect colonies. However, MacDonald *et al.* (1975) reported on two nest associates that apparently caused some damage to colonies.

Sphecophaga vesparum burra (Cresson) (Hymenoptera:Ichneumonidae), a pupal parasitoid, infested 33 of 39 *V. atropilosa* nests collected during 1971–1973 in southeastern Washington (Figure 3.8). It was suggested that destruction of the potential worker force in incipient colonies could seriously affect colony development. This parasitoid also commonly occurs in colonies of another *V. rufa* group species, *V. acadica* (Roush and Akre, 1978). This parasitoid also invades the nests

FIGURE 3.8. Adult female of *Sphecophaga vesparum burra* that has just emerged from a pupal cell in a *Vespla atropilosa* colony.

of *D. arenaria, D. maculata* (Greene *et al.*, 1976; Roush and Akre, 1978), *V. pensylvanica*, and *V. vulgaris* (Akre *et al.*, 1981; Roush and Akre, 1978), but rates of parasitism were low in these species and the ichneumonid did not seem to be an adverse factor in colony development.

Perhaps the high rates of infestation by *Sphecophaga* in nests of *V. atropilosa* and *V. acadica* can be best explained by their behavior toward these associates. In colonies of *V. atropilosa* and *V. acadica*, emerging *S. v. burra* are frequently antennated and then ignored. After emergence they may be attacked and chased by only one or two workers. Sometimes the parasites are killed, but more often they escape. Parasites introduced into nests of *V. atropilosa* in glass-bottomed observaton boxes (Akre *et al.*, 1976) did not cause widespread colony alarm. They usually entered the nest and walked on the combs before being attacked and chased away. Eventually a few were caught and killed. Conversely, *S. v. burra* introduced into nest boxes of *V. pensylvanica* caused widespread disturbance. Once a worker encountered a parasite, it immediately chased the parasitoid and soon 5 to 15 workers were agitated. The parasite was soon killed and sometimes even fed to the larvae as also occurs in *V. vulgaris* (Carmean *et al.*, 1981). More frequently they were taken outside and dropped. Subsequent invasions of this ichneumonid are probably vigorously resisted by the numerous guards at the entrance hole. The low rate of parasitism in *D. arenaria* and *D. maculata* (Greene *et al.*, 1976) colonies would seem to indicate these species also effectively defend the nest against invasion or conversely, are not preferred hosts of *S. v. burra*.

Triphleba lugubris (Meigen) (Diptera:Phoridae) was also reported to influence colony productivity in *V. pensylvanica* by destroying developing queen pupae. These flies are ordinarily scavengers and appear only when the colony has entered decline. The limited observations of these adult flies indicate they are generally ignored by the workers.

Another parasitoid, *Bareogonalis canadensis* (Harrington) (Hymenoptera:Trigonalidae), is rarely found in nests of *D. arenaria, V. acadica, V. pensylvanica*, and *V. vulgaris* (Carmean *et al.*, 1981). The workers chased or forcibly removed this parasitoid from *V. vulgaris* colonies, but they were not stung and usually escaped unharmed.

These few observations and the relatively low parasitism rate of vespine colonies (Spradbery, 1973) suggests that nest defense against inquilines is very effective. Indeed, biological control agents for pestiferous yellowjackets are frequently proposed, but no effective agent exists, and none is forseen (MacDonald *et al.*, 1976).

MICROBIAL DEFENSE

Although vespine colonies must succumb to diseases, no studies have been done indicating pathogenic microorganisms are involved. Spradbery (1973) reported that in some colonies larvae and pupae turned black and rapidly decayed, suggesting that a virus or bacteria may have been involved. Akre and Reed (1981b) also reported on rapid colony demise of *V. vulgaris* colonies in northern Idaho during wet weather and suggested a pathogen may have been responsible. There are no reports of pathogenic fungi although a fungus has been described from a nest of *Vespula lewisii* (Cameron) (Sagara and Kobayashi, 1979) and another fungus was reported infesting the envelope paper of an aerial yellowjacket (*D. arenaria*) (Durrell, 1965). Rather than being pathogenic, Durrell suggested that the fungal hyphae gave the envelope added strength. Other fungi are found on vespines, but it is not certain whether they attack the living insects or become established on dead ones (Edwards, 1980).

Norepinephrine has been found in the nest envelope of *V. vulgaris* (Bourden *et al.*, 1975) and *V. germanica* (Lecomte *et al.*, 1976). Although these authors did not speculate on possible functions of this material, it may serve as a protection against microbial invaders or fungi.

REFERENCES

Akre, R. D. (1982). The social wasps. In *Social Insects* (H. R. Hermann, ed)., Vol. 4, Chapter 1, pp. 1–105. Academic Press, New York.

Akre, R. D., and H. G. Davis. (1978). Biology and pest-status of venomous wasps. *Annu. Rev. Entomol.* 23, 19–42.

Akre, R. D., W. B. Garnett, J. F. MacDonald, A. Greene, and P. Landolt. (1976). Behavior and colony development of *Vespula pensylvanica* and *V. atropilosa* (Hymenoptera:Vespidae). *J. Kans. Entomol. Soc.* 49, 63–84.

Akre, R. D., A. Greene, J. F. MacDonald, P. J. Landolt, and H. G. Davis. (1981). The yellowjackets of America north of Mexico. *USDA Agric. Handbook* No. 552.

Akre, R. D., and H. C. Reed. (1981a). A polygynous colony of *Vespula pensylvanica* (Saussure) (Hymenoptera:Vespidae). *Entomol. News* 92, 27–31.

Akre, R. D., and H. C. Reed (1981b). Population cycles of yellowjackets in the Pacific Northwest (Hymenoptera:Vespinae). *Environ. Entomol.* 10, 267–274.

Akre, R. D., H. C. Reed, and P. J. Landolt. Nesting biology and behavior of the blackjacket, *Vespula consobrina* (Hymenoptera:Vespidae). *J. Kans. Entomol. Soc.* 55:373–405.

Akre, R. D., C. F. Roush, and P. J. Landolt. (1977). A *Vespula pensylvanica/ Vespula vulgaris* nest (Hymenoptera:Vespidae). *Environ. Entomol.* 6, 525–526.

Balduf, W. V. (1968). On the life of *Vespula vulgaris* (L.) and *V. maculifrons* (Buysson) (Hymenoptera:Vespidae). *Proc. Entomol. Soc. Wash.* 70, 332–338.

Bequaert, J. (1931). A tentative synopsis of the hornets and yellowjackets of America. *Entomol. Am.* 12, 71–138.

Bigelow, N. K. (1922). Insect food of the Black bear (*Ursus americanus*). *Can. Entomol.* 54, 49–50.

Birkhead, T. R. (1974). Predation by birds on social wasps. *Brit. Birds* 67, 221–229.

Bourdon, V., J. Lecomte, J. Damas, and J. Magis. (1975). Communication brève sur la présence de noradrenaline conjuguée dans la cire des rayons d'*Apis mellifera* Linné. *Arch. Int. Physiol. Biochim.* 85, 945–947.

Brian, M. V., and A. D. Brian. (1948). Nest construction by queens of *Vespula sylvestris* Scop. (Hymenoptera:Vespidae). *Entomol. Mon. Mag.* 84, 193–198.

Buckell, E. R., and G. T. Spencer. (1950). The social wasps (Vespidae) of British Columbia. *Proc. Entomol. Soc. Brit. Columbia* 46, 33–40.

Butler, C. G., D. J. C. Fletcher, and D. Walter. (1969). Nest-entrance marking with pheromones by the honey bee *Apis mellifera* L. and a wasp *Vespula vulgaris* L. *Anim. Behav.* 17, 142–147.

Carmean, D., R. D. Akre, R. S. Zack, and H. C. Reed. (1981). Notes on the yellowjacket parasite *Bareogonalis canadenis* (Hymenoptera:Trigonalidae). *Entomol. News* 92, 23–26.

Cederberg, B. (1979). Odour guided host selection in *Psithyrus. Entomol. Tidskr* 100, 128–129.

Chan, K. L. (1972). The hornets of Singapore: Their identification, biology, and control. *Singapore Med. J.* 13, 178–187.

Cobb, F. K. (1979). Honey Buzzard at wasps' nest. *Brit. Birds* 72, 59–64.

Davis, H. G. (1978). Yellowjacket wasps in urban environments. In *Perspectives in Urban Entomology* (G. W. Frankie and C. S. Koehler, eds.), XV Int. Congress on Entomology, 1976, pp. 163–185, Academic Press, New York.

DeJong, D. (1978). Insects: Hymenoptera (Ants, bees, and wasps). In *Honey Bee Pests, Predators and Diseases* (R. A. Morse, ed.), Chapter 9, pp. 139–175. Cornell Univ. Press, Ithaca, N.Y.

DeJong, D. (1979). Social wasps, enemies of honey bees. *Am. Bee J.* 119, 505–507, 529.

Duncan, C. P. (1939). A contribution to the biology of North American vespine wasps. *Stanford Univ. Publ. Biol. Sci.* 8, 1–271.

Durrell, L. W. (1965). Fungi in nests of paper wasps. *Am. Midl. Nat.* 73, 501–503.

Ebeling, Walter. (1975). *Urban Entomology*. Univ. Calif. Div. Agric. Sciences: Berkley.

Edwards, R. (1980). *Social Wasps: Their Biology and Control*. Rentokil, Sussex, England.

Evans, H. E. (1975). Social parasitism of a common yellowjacket. *Insect World Digest* 2, 6–13.

Francke, W., G. Hindorf, and W. Reith. (1978). Methyl-1, 6-dioxaspiro (4,5) decanes as odors of *Paravespula vulgaris* (L.) Angew. *Chem. Int. Ed. Engl.* 17, 862.

Free, J. B. (1970). The behavior of wasps (*Vespula germanica* L. and *V. vulgaris* L.) when foraging. *Insectes Soc.* 17, 11–20.

Gabriel, K. R., and D. C. Gilbert. (1975). Swallows feeding wasps to young. *Brit. Birds* 68, 248.

Gaul, A. T. (1941). Biology of the northeastern hornets. *Bull. Brooklyn Entomol. Soc.* 36, 38–41.

Gaul, A. T. (1953). Additions to vespine biology XI. Defense flight. *Bull. Brookyn Entomol. Soc.* 48, 35–37.

Greene, A., R. D. Akre, and P. J. Landolt. (1976). The aerial yellowjacket, *Dolichovespula arenaria* (Fab.): Nesting biology, reproductive production and behavior (Hymenoptera:Vespidae). *Melanderia* 26, 1–34.

Greene, A., R. D. Akre, and P. J. Landolt. (1978). Behavior of the yellowjacket social parasite, *Dolichovespula arctica* (Rohwer) (Hymenoptera: Vespidae). *Melanderia* 29, 1–28.

Harwood, R. F., and M. T. James. (1979). *Entomology in Human and Animal Health*, 7th ed. Macmillan, New York.

Hermann, H. R. (1971). Sting autotomy, a defensive mechanism in certain social Hymenoptera. *Insectes Soc.* 18, 111–120.

Hermann, H. R., and T. F. Dirks. (1974). Sternal glands in polistine wasps: Morphology and associated behavior. *J. Ga. Entomol. Soc.* 9, 1–8.

Horne, C. (1870). Notes on the habits of some hymenopterous insects from the northwest provinces of India with an appendix, containing descriptions of some new species of Apidae and Vespidae collected by Frederick Smith of the British Museum. *Trans. Zool. Soc. London* 7, 161–196.

Howell, J. O. (1973). Notes on yellowjackets as a food source for the bald faced hornet, *Vespula maculata* (L.). *Entomol. News* 84, 141–142.

Ishay, J. (1977). Acoustical communication in wasp colonies (Vespinae). *Proc. XV Int. Congress on Entomology (Washington, D.C., 1976)*, pp. 406–435.

Ishay, J., H. Bylinski-Salz, and A. Shulov. (1967). Contributions to the bionomics of the Oriental hornet (*Vespa orientalis* Fab.). *Isr. J. Entomol.* 2, 45–106.

Ishay, J., and A. Schwartz. (1973). Acoustical communication between the members of the Oriental hornet colony. *J. Acoust. Soc. Am.* 53, 640–649.

Jacobson, R. S., R. W. Matthews, and J. F. MacDonald. (1978). A systematic study of the *Vespula vulgaris* group with a description of a new yellowjacket species in eastern North America (Hymenoptera: Vespidae). *Ann. Entomol. Soc. Amer.* 71, 299–312.

Jeanne, R. L. (1970). Chemical defense of brood by a social wasp. *Science* 168, 1465–1466.

Jeanne, R. L. (1975). The adaptiveness of social wasp nest architecture. *Q. Rev. Biol.* 50, 267–287.

Jeanne, R. (1977). Behavior of the obligate social parasite *Vespula arctica* (Hymenoptera: Vespidae). *J. Kans. Entomol. Soc.* 50, 541–557.

Lecomte, J., V. Bourdon, J. Damas, M. Leclercq, and J. Leclercq. (1976). Présence de noradrenaline conjuguée dans les pariois du nid de *Vespula germanica* Linné. *C. R. Soc. Biol.* 170, 212–215.

Lord, W. D., D. A. Nicholson, and R. R. Roth. (1977). Foraging behavior and colony drift in *Vespula maculifrons*. *N. Y. Entomol. Soc.* 85, 186.

MacDonald, J. (1977). Comparative and adaptive aspects of vespine nest construction. *Proc. 8th Int. Congr. IUSSI, (Wageningen, The Netherlands)*, pp. 169–172.

MacDonald, J. F., R. D. Akre, and W. B. Hill. (1974). Comparative biology and behavior of *Vespula atropilosa* and *V. pensylvanica* (Hymenoptera: Vespidae). *Melanderia* 18, 1–66.

MacDonald, J. R., R. D. Akre, and W. B. Hill. (1975). Nest associates of *Vespula atropilosa* and *V. pensylvanica* southeastern Washington State (Hymenoptera: Vespidae). *J. Kans. Entomol. Soc.* 48, 53–63.

MacDonald, J. F., R. D. Akre, and R. W. Matthews. (1976). Evaluation of yellowjacket abatement in the United States. *Bull. Entomol. Soc. Am.* 22, 397–401.

MacDonald, J. F., and R. W. Matthews. (1975). *Vespula squamosa*: A yellowjacket wasp evolving toward parasitism. *Science* 190, 1003–1004.

MacDonald, J. F., and R. W. Matthews. (1981). Nesting biology of the eastern yellowjacket, *Vespula maculifrons* (Hymenoptera:Vespidae). *J. Kans. Entomol. Soc.* 54, 433–457.

MacDonald, J. F., R. W. Matthews, and R. S. Jacobson. (1980). Nesting biology of the yellowjacket, *Vespula flavopilosa. J. Kans. Entomol. Soc.* 53, 448–458.

Maschwitz, U. (1964a). Gefahrenalarmstoffe und Gefahrenalarmierung bei sozialen Hymenopternen. *Z. Vergl. Physiol.* 47, 596–655.

Maschwitz, U. W. (1964b). Alarm substances and alarm behavior in social Hymenoptera. *Nature* 204, 324–327.

Maschwitz, U. W. J., and W. Kloft. (1971). Morphology and function of the venom apparatus of insects—Bees, wasps, ants, and caterpillars. In *Venomous Animals and Their Venoms* (W. Bucherl and E. Buckley, eds.), Vol. III, *Venomous Invertebrates*, Chapter 44, pp. 1–60. Academic Press, New York.

Matthews, R. W., and J. W. Matthews. (1979). War of the yellowjacket queens. *Nat. Hist.* 88, 56–65.

Matsuura, M. (1970). Intraspecific invasion behaviour of *Vespa crabro flavofasciata* Cameron on early stage of nesting. *Life Study (Fukui)* 14, 21–26.

Matsuura, M. (1973). Nesting habits of several species of the genus *Vespa* in Formosa. *Kontyû* 41, 286–293.

Matsuura, M., and S. F. Sakagami. (1973). A bionomic sketch of the giant hornet, *Vespa mandarinia*, a serious pest for Japanese apiculture. *J. Fac. Sci. Hokkaido Univ. (Series 6, Zool.)* 19, 125–162.

Michener, C. D. (1975). The Brazilian bee problem. *Annu. Rev. Entomol.* 20, 399–416.

Miller, C. D. F. (1961). Taxonomy and distribution of nearctic *Vespula. Can. Entomol.* (Suppl. 22).

Nixon, G. E. J. (1935). Notes on wasps. IV. *Entomol. Mon. Mag.* 71, 106–111.

Ohtani, T., and T. Kamada. (1980). Worker piping: The piping sounds produced by laying and guarding worker honey bees. *J. Apicult. Res.* 19, 154–163.

Pack-Beresford, D. R. (1902). The nesting habits of *Vespa rufa. Irish Nat.* 11, 94–95.

Pagden, H. T. (1958). Some Malayan social wasps. *Malayan Nat. J.* 12, 131–148.

Piccioli, M. T. Marino, and L. Pardi. (1978). Studies on the biology of *Belonogaster* (Hymenoptera:Vespidae). *Ital. J. Zool.* 29, 179–228.

Potter, N. B. (1964). A study of the biology of the common wasp, *Vespula vulgaris* L., with special reference to the foraging behaviour. Ph.D. Dissertation, Univ. of Bristol, England.

Preiss, F. J. (1967). Nest site selection, microenvironment and predation of yellowjacket wasps, *Vespula maculifrons* (Buysson), (Hymenoptera:Vespidae) in a deciduous Delaware woodlot. M.S. Thesis, Univ. of Delaware.

Rau, P. (1929). The nesting habits of the bald-faced hornet, *Vespula maculata*. *Ann. Entomol. Soc. Am.* 22, 659–675.

Reed, H. C. (1981). Behavior and biology of the forest yellowjacket, *Vespula acadica*, and the obligate social parasite, *Vespula austriaca*. Ph.D. Dissertation Washington State Univ.

Reed, H. C. (1982). Biology and behavior of the forest yellowjacket, *Vespula acadica* (Sladen) and the obligate social parasite *Vespula austriaca* (Panzer) (Hymenoptera:Vespidae) Ph.D. Dissertation. Washington State University, Pullman.

Reed, H. C., and R. D. Akre. (1980). Behavior of the yellowjacket social parasite, *Vespula austriaca* (Panzer). *Proc. Wash. State Entomol. Soc.* 42, 584–585.

Reed, H. C., and R. D. Akre. 1983(a). Nesting biology of a forest yellowjacket, *Vespula acadica* (Sladen) (Hymenoptera:Vespidae), in the Pacific Northwest. *Ann. Ent. Soc. Am.* 76:582–590.

Reed, H. C., and R. D. Akre. 1983(b). Comparative colony behavior of the forest yellowjacket, *Vespula acadica* (Sladen) (Hymenoptera:Vespidae). *J. Kans. Entomol. Soc.* 56:581–606.

Reed, H. C., and R. D. Akre. 1983(c). Usurpation behavior of the yellowjacket social parasite, *Vespula austriaca* (Panzer) (Hymenoptera:Vespidae). *Am. Midl. Nat.* 110:419–432.

Reed, H. C., R. D. Akre, and W. B. Garnett. (1979). A North American host of the yellowjacket social parasite *Vespula austriaca* (Panzer) (Hymenoptera:Vespidae). *Entomol. News* 90, 110–113.

Richards, K. W. (1978). Nest site selection by bumble bees (Hymenoptera: Apidae) in Southern Alberta. *Can. Entomol.* 110, 301–318.

Richards, O. W. (1978). *The Social Wasps of the Americas Excluding the Vespinae.* British Museum of Natural History, London.

Roush, C. F., and R. D. Akre. (1978). Nesting biologies and seasonal occurrence of yellowjackets in northeastern Oregon forests (Hymenoptera:Vespidae). *Melanderia* 30, 57–94.

Sagara, N., and T. Kobayashi. (1979). A fungus procured from abandoned nests of *Vespula lewisii. The Insectarium Tilyo Zool. Park Soc.* 16, 8–11.

Sakagami, S. F., and K. Fukushima. (1957). *Vespa dybowskii* Andre as a facultative temporary social parasite. *Insectes Soc.* 4, 1–12.

Saslavasky, H., J. Ishay, and R. Ikan. (1973). Alarm substances as toxicants of the Oriental hornet, *Vespa orientalis. Life Sci.* 12, 135–144.

Schmidt, J. O., and M. S. Blum. (1979). Toxicity of *Dolichovespula maculata* venom. *Toxicon* 17, 645–648.

Schmidt, J. O., M. S. Blum, and W. L. Overal. (1980). Comparative lethality of venoms from stinging Hymenoptera. *Toxicon* 18, 469–474.

Schmidt, J. O., H. C. Reed, and R. D. Akre. (1984). Venoms of a parasitic and two nonparasitic species of yellowjackets (Hymenoptera:Vespidae). *J. Kans. Entomol. Soc.* 57:[in press].

Sharp, J. L., and J. James. (1979). Color preference of *Vespula squamosa. Environ. Entomol.* 8, 708–710.

Spradbery, J. P. (1973). *Wasps: An Account of the Biology and Natural History of Solitary and Social Wasps.* Univ. Washington Press, Seattle.

Taylor, L. H. (1939). Observations on social parasitism in the genus *Vespula* Thomson. *Ann. Entomol. Soc. Am.* 32, 304–315.

Turillazzi, S., and A. Ugolini. (1979). Rubbing behavior in some European *Polistes. Monit. Zool. Ital.* 13, 129–142.

van der Vecht, J. (1957). The vespine of the Indo-Malayan and Papuan areas (Hymenoptera:Vespidae). *Zool. Verh. Rijkmus Nat. Hist. (Leiden)* 34 1–83.

Wilson, E. O. (1971). *The Insect Societies.* Belknap Press of Harvard Univ. Press, Cambridge, Mass.

Yamane, S. (1975). Taxonomic notes on the subgenus *Boreovespula* Blüthgen (Hymenoptera:Vespidae) of Japan, with notes on specimens from Sakhalin. *Kontyû* 43, 343–355.

Yamane, S. (1976). Morphological and taxonomic studies on vespine larvae, with reference to the phylogeny of the subfamily Vespinae (Hymenoptera:Vespidae). *Insecta Matsumurana* (NS) 8, 1–45.

Yamane, S., and S. Makino. (1977). Bionomics of *Vespa analis insularis* and *V. mandarinia latilineata* in Hokkaido, northern Japan, with notes on vespine embryo nests (Hymenoptera:Vespidae). *Insecta Matsumurana* (NS) 12, 1–33.

Yamane, S., R. E. Wagner, and S. Yamane. (1980). A tentative revision of the subgenus *Paravespula* of Eastern Asia (Hymenoptera:Vespidae). *Insecta Matsumurana* (NS) 19, 1–46.

4

Defensive Behavior and Defensive Mechanisms in Ants

Alfred Buschinger and Ulrich Maschwitz

INTRODUCTION

Defense—in the widest sense—may comprise a variety of mechanisms that serve to maintain a certain homeostasis for an individual, or for a society, in the case of social insects.

Defensive mechanisms, therefore, are all morphological, chemical, and behavioral tools that help to preserve the life and soundness of the animal, mechanisms that lower the impact not only of aggressors, competitors, parasites, and predators, but also of microorganisms. Defensive behavior in the social insects may be defined as "any behavioral display that offers protection or potential protection to the individual or the individual's colony" (Hermann and Blum, 1981).

Defense may consist of such varying mechanisms as a tough cuticle, camouflage, aposematism, flight, withdrawal into shelters or the soil for hibernation, storage of food reserves, chemical protection against microorganisms, and other, more or less indirect or "primary mechanisms" (Robinson, 1969), as well as direct combat against enemies, with the aid of mechanical or chemical weapons. For a detailed survey see Edmunds (1974). This review is concerned mainly with these direct mechanisms of defense in ants; however, we do not completely omit the indirect, protective methods. We only defer most of the mechanisms of protection against abiotic impacts.

The family Formicidae comprises about 10,000 species, all of which live in eusocial communities. A survey of ant systematics is given in Table 4.1. Like other social insects, bees, wasps, and termites, they often form huge accumulations of adults and brood stages,

TABLE 4.1. Subfamilies, Genera, Subgenera, and Species of Ants Mentioned in the Text

Subfamily	Genus	Species
Myrmeciinae	*Myrmecia*	—
Nothomyrmeciinae[a]	—	—
Ponerinae	*Amblyopone*	*australis*
	Anochetus[b]	—
	Brachyponera	*senaarensis*
	Dinoponera	—
	Harpegnathus	*saltator*
	Leptogenys	*ocellifera*
	Leptogenys	*chinensis*
	Megaponera	*foetens*
	Odontomachus	—
	Pachycondyla	*tridentata*
	Paltothyreus	*tarsatus*
	Paraponera	*clavata*
	Ponera	—
Ecitoninae	*Eciton*	—
Leptanillinae	—	—
Dorylinae	*Aenictus*	*fergusoni*
	Anomma	—
	Dorylus	—
Pseudomyrmecinae	*Pachysima*[c] (= *Tetraponera*)	—
	Pseudomyrmex	*ferruginea*
	Tetraponera	—
Myrmicinae	*Acromyrmex*	*versicolor*
	Aphaenogaster	*fulva*
	Atta	*cephalotes*
	Atta	*sexdens*
	Cardiocondyla	—
	Carebara	—
	Cataulacus	—
	Chalepoxenus[c](= *Leptothorax*)	—
	Cremastogaster (*Physocrema*)	*difformis*
	Cremastogaster (*Physocrema*)	*inflata*
	Cremastogaster (*Physocrema*)	*scutellaris*
	Cryptocerus[c] (= *Cephalotes*)	—
	Cyphomyrmex	—
	Diplorhoptrum[c] (= *Solenopsis*)	*banyulensis*
	Diplorhoptrum	*fugax*
	Harpagoxenus	*sublaevis*
	Leptothorax	*acervorum*
	Leptothorax	*curvispinosus*
	Manica	*rubida*
	Meranoplus	*bicolor*

TABLE 4.1. *(continued)*

Subfamily	Genus	Species
	Messor	*barbarus*
	Messor	*meridionalis*
	Moellerius[c] (*Acromyrmex*)	—
	Monomorium	*minimum*
	Monomorium	*pharaonis*
	Mycocepurus	—
	Myrmecina	—
	Myrmica	*rubra*
	Myrmica	*scabrinodis*
	Myrmica	*schencki*
	Myrmicaria	*eumenoides*
	Novomessor[c] (= *Aphaenogaster*)	—
	Paracryptocerus[c] (= *Cephalotes*)	—
	Pheidole	*biconstricta*
	Pheidole	*dentata*
	Pheidole	*embolopyx*
	Pheidole	*fallax*
	Pheidole	*lamia*
	Pheidole	*militicida*
	Pheidole	*pallidula*
	Pheidole	*sykesii*
	Pogonomyrmex	*badius*
	Pogonomyrmex	*barbatus*
	Pogonomyrmex	*comanche*
	Pogonomyrmex	*maricopa*
	Rogeria	*stigmatica*
	Solenopsis	*geminata*
	Solenopsis	*invicta*
	Solenopsis	*saevissima*
	Stenamma	—
	Strongylognathus	—
	Strumigenys	—
	Tatuidris	—
	Tetramorium	*caespitum*
	Tetramorium	*simillimum*
	Trachymyrmex	—
	Xenomyrmex	*floridanus*
	Zacryptocerus[c] (= *Cephalotes*?)	*multispinosus biguttatus*
	Zacryptocerus	*multispinus*
	Zacryptocerus	*umbraculatus*
	Zacryptocerus	*varians*
Aneuretinae[a]	*Aneuretus*	*simoni*

(continued on next page)

TABLE 4.1. *(continued)*

Subfamily	Genus	Species
Dolichoderinae	*Bothriomyrmex*	—
	Conomyrma[c] (= *Dorymyrmex*?)	*bicolor*
	Dolichoderus (*Hypoclinea*)	—
	Iridomyrmex	*humilis*
	Paratrechina	*longicornis*
	Tapinoma	*erraticum*
	Tapinoma	*melanocephalum*
	Technomyrmex	*albipes*
Formicinae	*Acanthomyops*	*claviger*
	Calomyrmex	—
	Camponotus	*sericeus*
	Camponotus (*Colobopsis*)	*saundersi*
	Camponotus (*Colobopsis*)	*truncatus*
	Cataglyphis	*bicolor*
	Cataglyphis	*cursor*
	Chthonolasius[c] (= *Lasius*)	—
	Colobopsis[c] (= *Camponotus*)	—
	Dendrolasius[c] (= *Lasius*)	—
	Formica	*lugubris*
	Formica	*polyctena*
	Formica	*yessensis*
	Formica (*Serviformica*)	*fusca*
	Formica (*Raptiformica*)	*pergandei*
	Formica (*Raptiformica*)	*subintegra*
	Gigantiops	—
	Lasius	*americanus*
	Lasius	*niger*
	Lasius (*Chthonolasius*)	*flavus*
	Lasius (*Chthonolasius*)	*umbratus*
	Lasius (*Dendrolasius*)	*fuliginosus*
	Myrmecocystus	*mimicus*
	Oecophylla	*longinoda*
	Oecophylla	*smaragdina*
	Polyergus	—
	Polyrhachis	—
	Proformica	*epinotalis*
	Rossomyrmex	*proformicarium*

[a] After Taylor (1978).
[b] After Brown (1976).
[c] After Brown (1973).

usually bound to a certain locality, a nest. Unlike many wasp and bee colonies, however, the societies of ants in general are long lived, often reaching an age of several decennia, and thus accessible to various kinds of aggressors throughout the year, and with a rather predictable continuity.

Ant colonies, with their internal nest structure, sometimes with temperature regulation (*Formica*, e.g., Steiner 1924, 1929; Kneitz, 1970), protection against climatic impacts and aggressors, and food and organic refuse resources, often form real "ecosystems" with plant lice sucking in plants or roots and providing the ants with honeydew, with fungi and other decomposers, and with numerous kinds of so-called guests, many of which have developed toward parasites and predators of ant broods.

These facts mainly imply the risk that predators, parasites and excitants of diseases may specialize on ants or certain ant species groups. Consequently, a variety of defensive mechanisms against such menaces, ranging from fungi and bacteria to nematodes, cestodes, mites and spiders, beetles, ichneumonids, flies, and other insects to amphibians, reptiles, birds, and mammals, have been developed in the Formicidae.

However, as many authors have emphasized, ants themselves are the most important enemies of ants, with various kinds of inter-specific and intraspecific impact on each other (e.g., Maidl, 1934). Competition for food and nest sites, territorial interactions, and predation on other ant species occur as well as parasitism, for example, as in temporary parasites, inquilines, and slavemaker ants. Defense against other ants, therefore, plays a major role in the protective patterns of ant behavior.

The necessity of defense, not only of the individual, as in solitary insects, but of the society, with its vulnerable queens, callows, and brood stages, may be a further cause for the development of highly effective fighting mechanisms. Social life with caste polymorphism and polyethism, on the other hand, provides the chance to produce, in addition to ordinary workers, soldiers of various kinds, who may be specialized for fight and defense up to the point of self-sacrifice (*Camponotus saundersi*, Maschwitz and Maschwitz, 1974; *Pogonomyrmex* spp., Hermann, 1971).

The defensive potential of ants may be illustrated by the remarkable observations of neotropical wasp species of the genera *Polybia* and *Protopolybia*, which nest exclusively in trees inhabited by certain dolichoderine ants. These ants do not prey on the wasps but present a barrier to other predators, of the ants as well as of the wasps. Also birds such as orioles gain protection from the ants by nesting in the same trees (Evans and Eberhard, 1970; Wheeler *et al.*, 1975; Howse, 1975).

However, as is shown in the following sections, by far not all fighting mechanisms in ants are purely defensive. Often we find mechanisms that may serve in both aggression and defense, such as the stinging apparatus, poison secretion, or biting mandibles. This may be due to the fact that many ant species themselves are predators or depend at least in part on other arthropods for prey. For example, in territorial encounters, one of the competing colonies is the aggressor and the other the defender. In this case the fighting mechanisms for offense and defense are often the same. A clear distinction between aggressive and defensive mechanisms and behaviors, therefore, is often impossible. On the other hand, ants are exposed to such a variety of menaces and dangers that specialization in defense against one or a few of them often seems to offer no advantage.

In the following section, we first present a survey of mechanisms, structural as well as chemical and behavioral, involved in the defense of ants. Further sections are devoted to the organization of social defense and the evolutionary aspects of the development of defensive mechanisms.

MORPHOLOGICAL STRUCTURES, CHEMICAL SECRETIONS, AND BEHAVIORS INVOLVED IN SELF-PROTECTION AND DEFENSE OF ANTS

Defensive mechanisms in ants are generally restricted, with few exceptions, to the sterile caste, to workers and soldiers if present. Males lack a stinging apparatus, a structure derived from the hymenopteran female ovipositor, and we do not know of any defensive behavior in ant males apart from hiding or flying off when the nest is disturbed. Whether their strongly smelling mandibular gland secretions may be used as repellent substances is not yet clear (Blum and Hermann, 1978a, see following). Similarly, as in other social insects, the queens generally prefer to withdraw into the nest in dangerous situations (Maidl, 1934). Little is known about defensive mechanisms in brood stages, larvae, or pupae. They are completely dependent from the care of workers that carry the broods to places in the nest with suitable climatic conditions, rescue the immatures in case of peril, and lick them continuously in order to prevent growth of microorganisms on their cuticle (Wheeler, 1910). Wheeler and Wheeler (1979) claim that the hairy tubercles of ponerine larvae might afford protection against cannibalism through sister larvae; however, experimental proof is lacking, and the tubercles are not present in all Ponerinae. The formation of a pupal cocoon in most subfamilies, with the exception of Myrmicinae and Dolichoderinae, provides protection against desiccation, bacteria, or fungal spores (Schmidt, 1961), as well as the attack of mites.

A tough and hard integument is another widespread protective measure in arthropods. In ants, we find both heavily armored species and others with a very thin and pliable cuticle. It is difficult to find clear correlations between chitinization and life habits or systematic groups, and different strategies seem to compete or to conflict with each other. Wheeler (1910) claims that the chitinous investment is hard and brittle in many of the more primitive groups, especially Ponerinae, and thinner in the more recently developed forms (most Dolichoderinae and Formicinae). But *Polyrhachis* (Formicinae) and *Dolichoderus* (Dolichoderinae), which are heavily sclerotized, demonstrate that this relation is not generally consistent.

Species with a subterranean life, such as the subgenus *Chtonolasius* among the Formicinae, often have a thin cuticle, as one might expect. However, many Ponerines with hypogeic life habits are heavily armored. Ants with epigeic or arboreal life may have a hard and tough cuticle, like many other Ponerinae, *Cryptocerus, Atta, Cataulacus*, and *Polyrhachis* species; however, the opposite is true for numerous epigeic *Lasius, Formica*, and dolichoderine species. Some correlations seem to exist between armoring and the speed of movement. Quickly moving ants (e.g., many Dolichoderinae or Formicinae), with the option of rapid flight, need not and must not be strongly chitinized, since heavy armor impedes fast movement.

However, a further correlation seems to exist between the structure of the cuticle and the most important means of active defense. Stinging ants such as Myrmeciinae, Ponerinae, Pseudomyrmecinae, and many Myrmicinae, preferably are heavy armored, whereas the Formicinae and Dolichoderinae, which defend themselves by smearing the attackers with formic acid and other defensive substances, respectively, more often have thin cuticles. A tough cuticle has a clearly protective value against stinging and biting opponents, whereas the chemical properties of the cuticle are more important when aggressive substances are externally applied. With respect to the above-cited statement that ants themselves are their most important enemies, this correlation of an armoring and fighting system seems reasonable to a certain extent (see Evolutionary Aspects of Defensive Systems in Ants).

Among the heavy-armored ants, there are a number of genera with manifold spines and hooklike protuberances of the head, alitrunk, and petiole, which should have a protective value (Figure 4.1). Examples of the myrmicine genera *Atta, Trachymyrmex, Moellerius* and the formicine genus *Polyrhachis* are depicted by Wheeler (1910), Dumpert (1978), and others. Among the Dolichoderinae, the genus *Dolichoderus* includes species with spines (Wheeler, 1922a). These structures seem to be especially effective against vertebrate predators. In several species—for example, a yet undetermined *Polyrhachis* observed in the Indomalayan area—the spines hook into the skin to such a degree that it is difficult to shake off an ant that has been picked up with the fingers.

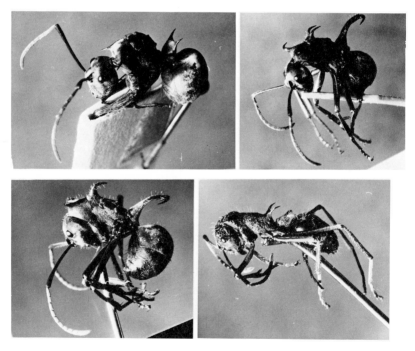

FIGURE 4.1. Spines and hooklike protuberances of the alitrunk and petiole as defensive structures in Southeast Asian *Polyrhachis* species.

In a number of genera, for example, *Cryptocerus, Cataulacus, Strumigenys* and *Meranoplus* (Wheeler, 1910) and *Tatuidris* (Brown and Kempf, 1967), the sensitive antennae can be concealed in deep grooves under broad ridges along the sides of the head. *Harpagoxenus* and *Chalepoxenus*, slavemaking myrmicines specialized for fighting during their slave raids, have similar antennal grooves.

The formation of a morphologically distinct subcaste, the soldiers, occurs in a number of ant species. In genera with a highly polymorphic worker caste, it is preferable that the largest specimens function as soldiers, as in army ants of the genera *Eciton* or *Dorylus* (Dorylinae). During a raid, their soldiers, with very large heads and long, sickle-shaped mandibles, walk or stand on both sides of the column and "serve as a defense force" (Schneirla, 1971; Wilson, 1971).

A higher specialization is represented by the soldiers of *Pheidole* species, who have disproportionately large heads, shearing or crushing mandibles, and a behavior distinct from that of the workers (for example, *Pheidole dentata, P. militicida*). They normally do not work much for the colony except during defense, and they do not forage (Creighton and Creighton, 1959; Wilson, 1975a; Kugler, 1978).

Much more intriguing is the structure and function of the soldier subcaste in certain genera such as *Camponotus* (*Colobopsis*), in *Pheidole lamia*, and in others; the plug-shaped heads of these soldiers are used to stopper the nest entrances (Wheeler, 1910; Brown, 1967). A worker ant returning to the nest has to palpate the soldier's head with its antennae; the soldier then retreats a little and gives way to the nest mate. In some instances, such as in *Colobopsis* (Figure 4.2), the queens also have peculiar pluglike heads. Closing of nest entrances with a specialized part of the body ("phragmosis") supposedly also occurs in *Pheidole embolopyx* (Brown, 1967); however, in this species the queen's gaster is formed like a plug.

Phragmosis, and plugging up the nest entrances by such specialized individuals, is clearly a protective measure for the society. It may lend itself to further protective and defensive behaviors.

Protective Behaviors of Ants

Closing of nest entrances with parts of the body was also observed in the formicine, *Proformica epinotalis* (Marikovsky, 1974). During the attack of a slavemaker ant, *Rossomyrmex proformicarium*, the defending *Proformica* not only stopped the entrances of their nest in the soil with small particles of sand and earth, but also with repletes, which stick in

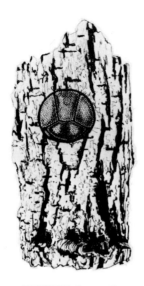 

FIGURE 4.2. *Camponotus* (*Colobopsis*) truncatus soldiers with plug-shaped heads stopper the next entrances. (Redrawn from Linsenmaier, 1972).

the narrow passages because of their inflated gasters. *Cremastogaster* species living in hollow acacia thorns are said to plug the only nest entrance with their gasters and to defend the nest against raiding *Anomma* using poison sprayed at the aggressors (Forel, 1948).

Closing of nest entrances with small stones or soil particles as a protective means against climatic impacts is a widespread behavior among ants. It may also function as a protective means against predators. *Harpegnathus saltator*, a tropical ponerine, as one example, closes the entrances every night at sunset and opens them again in the morning. This may protect against *Aenictus* species (Dorylinae), which usually hunt during the night, and exclusively for ants (U. Maschwitz, unpublished). Similar observations on the behavior of *Messor barbarus meridionalis* are reported by Forel (1948). Schneider (1971) suggests that *Cataglyphis bicolor* closes the entrances, especially to avoid the attacks of certain spiders (Zodariidae) that often lurk outside the nest. In areas where these spiders do not occur the entrances of the ant hills are said to remain open during the night.

The construction of nests in the soil, in wood or bark, in rock crevices, and so forth may be considered a protective behavior in the widest sense, that is, an anachoresis. Even ants "without nests," such as many army ant species, form clusters of workers during the night or during the stationary phase of their brood cycle, where most of the shelter for the queen and the broods is provided by the worker's bodies (Schneirla, 1971; Wilson, 1971). Furthermore, ants often excavate subterranean galleries from the nest to their food sources, or roof over the runways with sheds of plant material, soil, or wood particles, a behavior that may prevent predation as well as noxious climatic impacts (For a detailed description, see Wheeler, 1910; Wilson, 1971; Dumpert, 1978).

In a few species, the construction of special nest entrances, so-called entrance turrets, has been observed. The European *Myrmica schencki* forms a 1- to 2-cm-high tubular construction above the entrance to the nest in the soil. It is made from plant material (Heller, unpublished). A similar turret approximately 1 cm high was found in *Camponotus sericeus*, and *Meranoplus bicolor* forms a funnel-shaped structure from soil after rainfalls (U. Maschwitz, unpublished). The most complicated nest entrance structure is reported by Forel (1948) in *Pheidole sykesii* from India. It consists of an earthen calyx surrounded by up to eight concentric ramparts with a maximum diameter of 3 cm. The functions of these structures until now have not been investigated.

Flight is one of the most widespread reactions of ants when a nest is attacked or disturbed. Ordinarily the queens and young indoor workers, the latter carrying immatures, withdraw into the depths of the nest, whereas the older workers and, if present, soldiers, engage in active defense of the nest (see below). Not only fast running but also jumping over considerable distances occur during flight and attack in several species. Wheeler (1922b) differentiates between species that leap forward with their legs like leafhoppers (e.g., *Myrmecia, Gigantiops, Harpegnathus*), and others jumping backward with the aid of a click mechanism of their mandibles. Whether the latter is always "intended" as a means of sudden escape or merely as a by-product of a failed grasping movement may be questioned; however, the jumping reaction of *Harpegnathus saltator* is a definite flight reaction on disturbance (U. Maschwitz and M. Hahn, unpublished).

A peculiar deviation of ordinary flight behavior occurs in *Formica* (*Serviformica*) species during the attack of slavemaking ants of the genus *Polyergus*. Presumably alarmed by a pheromone of the slave-makers, the slave species workers try to escape from the nest, carrying brood stages, and climb up the surrounding vegetation (Forel, 1874;

FIGURE 4.3. Organized flight of an *Oecophylla* colony during an attack of *Technomyrmex*. The ants form a ball and then drop to the ground.

Wasmann, 1934). During raids of the ant-eating doryline, *Aenictus fergusoni* in Sri Lanka, the attacked myrmicine, *Meranoplus bicolor*, often exhibits the same behavior (U. Maschwitz, unpublished), and *Messor* species in North Africa fly to the surface of their nests when attacked by a subterranean raiding party of a *Dorylus* species. *Oecophylla smaragdina*, on the other hand, forms clusters of brood and workers outside the silk-woven leaf nest when raided by *Technomyrmex albipes* (Dolichoderinae). The clusters of thousands of ants finally drop to the ground (Figure 4.3). A comparable reaction is often observed in earth-dwelling species (e.g., *Myrmica, Lasius*) when their nesting site is slowly flooded; the societies then move to the top of grass stems or poles near the nest and form dense clusters around their broods and queens. In the tropics, army ants, leaf-cutting Attini, *Solenopsis geminata*, and *S. saevissima* were observed during inundations to form swimming balls of ants, which then eventually reach the bank again (Maidl, 1934).

Only a few ant species exhibit a thanatosis behavior when attacked; they roll themselves up and remain motionless for some time, "feigning death." Although this behavior has been observed in a variety of genera, such as *Myrmecina, Stenamma, Cyphomyrmex, Trachymyrmex, Mycocepurus* (Wheeler, 1910; Creighton, 1950; Schumacher and Whitford, 1974), *Tapinoma melanocephalum* (Goetsch, 1934), and *Cataulacus* (U. Maschwitz, unpublished), its protective value has never been experimentally investigated.

Finally, a few instances of specialized behaviors, defenses against distinct menaces, are mentioned here. Thus, about 200 species of socially parasitic ants with various relations to their host species have been described. Little is known about the defense of the host ants against parasitic queens who try to penetrate into the nest. Many of them are killed; however, they become victims of the general defense against ants of a different colony or species who get into the territory or near the nest. No special protective mechanisms against socially parasitic ants seem to occur. On the other hand, a few social parasites apparently have developed protective behaviors against the attacks of the host species. Thus, the temporary social parasitic *Lasius umbratus* queen first catches a worker of the host species *Lasius niger*, chews it, smears the pulp on her own body, and then penetrates into the host nest, apparently camouflaged with the nest odor of the colony (Crawley, 1909).

Eibl-Eibesfeldt and Eibl-Eibesfeldt (1968) describe a peculiar behavior of leaf-cutter ants (*Atta cephalotes*) that is interpreted as a specialized defense against a parasitic phorid fly (*Apocephalus*). When media workers are engaged in carrying or cutting a piece of leaf and thus cannot defend themselves, an accompanying minima worker, often riding on the back of the larger worker or crawling around on the piece of leaf, attacks the fly and prevents her from laying an egg on the ants.

Defensive Functions of the Mandibles

The mandibles of ants are often termed their "hands" because they are used for manifold activities, for example, digging in soil or gnawing wood during nest construction, transportation of nest material, brood stages, nestmates, refuse, and so forth, cleaning their own or their nest mates' bodies, and catching prey and cutting it in pieces. Last but not least, the mandibles are one of the major defensive and offensive weapons in many ants (Maschwitz, 1975).

Apparently, as an adaption to these varying requirements, the shape of the mandibles in most ants is rather uniform, more or less triangular and with a dentate masticatory border. A number of deviations, either as adaptations to special kinds of food, or to the function as a weapon, are depicted by Wheeler (1910). Notably among the Ponerinae the mandibles exhibit numerous structural peculiarities, apparently mostly due to food specialization (Wheeler, 1936). As examples, the genera *Odontomachus* and *Anochetus* are discussed here. Their long, linear mandibles are inserted closely together at their bases, and they can be kept widely open, at an angle of about 180°. As soon as the long hairs at the inner edge of the mandibles are touched by their prey, they are suddenly closed by means of a click mechanism and grasp the victim (Maidl, 1934). The extremely long mandibles of *Harpegnathus saltator* function like a pair of forceps that enable the ant to catch fast-moving prey objects such as cockroaches or bugs (Maschwitz *et al.*, 1979).

Ordinarily, the mandibles, even if the ant has other powerful fighting systems such as a stinger or poisonous secretions, have a supporting function during fight or defense. Many stinging Myrmicinae usually grasp a victim, or an aggressor, with the mandibles, and then bend the gaster forward and try to insert the stinger (Forel, 1920; Maidl, 1934). Even some formicines (e.g., *Formica*), who are able to spray an aggressor with formic acid over a distance, often bite and simultaneously discharge the poison at the opponent.

During encounters of mandible-fighting ants, the mandibles are used either to cut off the extremities of the adversary, or to keep the enemy at one of its legs or antennae. Especially when a group of ants pulls on several appendages of a victim, and in opposite directions, the victim is immobilized and often torn apart. This "gang pulling" (Kugler, 1978) not only occurs as a means to overcome large prey but is also very often encountered during fights between ants themselves (Figure 4.4).

Mandibles specialized for fighting have been developed mainly in soldiers of species with such a subcaste. Among the army ants the long, sickle-shaped mandibles of the so-called soldiers have already been mentioned. In *Pheidole*, their jaws are often toothless, with sharp, cutting edges. In leaf-cutting Attini, the larger individuals also have very sharp mandibles that serve mainly for cutting plant material; however, the largest specimens primarily function as soldiers (Kugler, 1978). As the authors themselves experienced, their mandibles are remarkably efficient in cutting the human skin.

Conspicuous specializations of mandibles for fighting with other ants are found in some slavemaker ants. The myrmicine genus *Harpagoxenus* has broad, toothless mandibles with sharp masticatory borders. They serve to cut off the extremities of slave species workers who defend their nests and broods (Buschinger *et al.*, 1980). In this

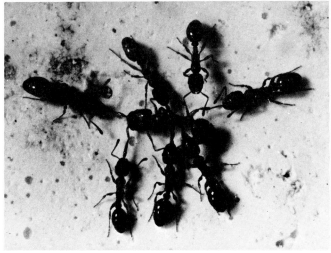

FIGURE 4.4. Gang pulling of *Leptothorax acervorum* against a *Harpagoxenus sublaevis* ergatomorphic queen that has attacked the *Leptothorax* colony.

genus (for example, *Harpagoxenus sublaevis*, Maschwitz and Kloft, 1971) the stinger is somewhat reduced and cannot be used for the injection of poison.

The formicine *Polyergus*, and the myrmicine *Strongylognathus*, both have long, saber-shaped mandibles. During fights with workers of raided slave species colonies these slavemakers sometimes pierce the heads of the defenders. Although the mandibles of these slavemakers are mainly offensive weapons, they must be mentioned in the context of defense as well; several authors have stressed that some such ants, for example, *Polyergus*, only rarely kill the defending ants, and that they make use of the piercing mandibles only in self-defense, when seized by one or several of the slave species workers (Wasmann, 1891; Dobrzańska, 1978).

The Stinger as a Defensive Weapon

Among the Aculeata *s.str.* to which the ants belong, the sting apparatus has lost its original ovipositor function, and has entirely become a fighting organ. The stinger, together with the mandibles, is the primary weapon of the Formicidae. However, it has retained this

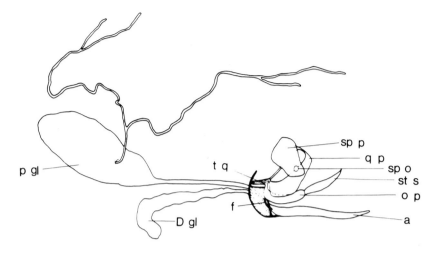

FIGURE 4.5. Sting apparatus of a *Diacamma* worker: a, aculeus; D gl, Dufour's gland; f, furcula; o p, oblong plate; p gl , poison gland; q p, quadrate plate; sp o, spiracular orifice; sp p, spiracular plate; st s, sting sheath. (Drawing by K. Jessen).

function only in a minority of the species. Frequently the ant sting apparatus was reduced and modified in its morphology and function (for reviews see Maschwitz and Kloft, 1971; Kugler, 1979a).

The stinger of the aculeata lies in a chamber formed by invagination of abdominal segments 8–10 into the seventh segment (abdominal sequent I = epinotum/propodeum). The modified sclerites of these segments together with their musculature and appendages form a cuticular injection apparatus with the actual poison gland and Dufour's gland adhering (Figure 4.5). It consists of a tubular injection instrument, the aculeus, and a basal system of muscle-bearing plates and arcs that serve in protrusion and injection of the sting into the opponent. In most Aculeata, the so-called pump stingers, through the stinging movements and with the aid of pumping membranes inside the aculeus, the secretions of the adhering glands are sucked in and ejected into the adversary. A muscular locking apparatus of the venom duct within the aculeus allows a dosage or throttling of secretion flow during stinging (Maschwitz and Kloft, 1971). *Sensilla campaniformia* in the aculeus and *sensilla trichodea* and *campaniformia* in a pair of sting sheaths facilitate the localization of a suitable stinging spot (Rathmayer, 1962; Hermann and Douglas, 1976).

Numerous recent investigations on the stinging apparatus in stinging ants of different groups revealed that the morphology of this apparatus corresponds to that of the typical stinger of the Aculeata (Blum and Hermann, 1978a). Surprisingly, during the past few years only a number of glandular complexes were found in the sting chamber and on the cuticular plates. These included: sting sheath glands, basal sting sheath glands, spiracular plate glands, and latero-ventral sting chamber glands (Hölldobler *et al.*, 1976; Jessen *et al.*, 1979). All of these open into the sting chamber. Supposedly these glands produce no secretions involved in fighting. However, with the exception of the sting sheath glands of *Atta* (Jaffé *et al.*, 1979; Bazire-Benazét and Zylberberg, 1979), their functions are not yet elucidated.

The poison gland of ants, in its typical form, consists of a pair of tubes secreting the venom, a saclike reservoir, and a canal that leads into the aculeus. This type of poison gland is represented in all subfamilies of ants investigated, with the exception of a highly modified organ in the Formicinae (for references, see Blum and Hermann, 1978a).

The venoms of the numerous ants capable of stinging have been investigated in only a small selection of species. Except in a few Myrmicinae the poison glands produce aqueous solutions of protein-aceous venoms, as is also typical for other hymenopterans (see reviews in Stumper, 1960; Maschwitz and Kloft, 1971; Blum and Hermann, 1978a). Until now there are exact analyses only of the venoms of *Myrmecia* (Blum and Hermann, 1978a), *Myrmica* (Jentsch, 1969), and *Pogonomyrmex* (Schmidt and Blum, 1978), where non-enzymatic and enzymatic proteins, polypeptides, and biogenic amines were found. Since the venoms contain mixtures of these substances, a wide range of pharmacological activities can be expected.

For humans the effects of sting venoms from various ant species are quite variable, but rarely dangerous as far as is known. The sting effects of many Ponerinae are painful but short lasting. Some large species, however, can cause long-lasting pain together with paralytic effects and fever (*Dinoponera, Paraponera clavata, Paltothyreus*; Stumper, 1960; Maschwitz and Kloft, 1971; Blum and Hermann, 1978a). Painfully stinging species are also found among the Myrmeciinae (*Myrmecia*), Pseudomyrmecinae (e.g. large *Tetraponera* species), and Myrmicinae (*Pogonomyrmex, Manica*).

Since the aqueous protein venoms develop their poisonous actions only within the body or on mucous membranes, they have to be injected into the opponent. If the adversary is a heavily sclerotized arthropod, the ants often bite and cling to it in order to have a support for stinging. The same behavior is sometimes observed when they sting a human enemy; however, some Ponerines (e.g., *Leptogenys*) are especially able to sting very quickly without biting.

Sting "autothysis," a specialized means of venom application that also occurs in other social Hymenoptera, is found in a number of species of the myrmicine genus *Pogonomyrmex*, for example, *P. comanche* and *P. maricopa* (Hermann, 1971). In these ants, especially when they sting into the tough skin of vertebrates, the stinging apparatus together with its ganglia and glands remains lodged in the wound due to long barbs on the stinger. Whether further morphological modifications, as seen in the honeybee (Rietschel, 1937), facilitate the extraction of the sting apparatus is not yet known. Whereas an opponent can rather easily strip off a complete stinging ant, this is nearly impossible with the small and firmly caught-in sting apparatus alone, which then works on injecting its complete venom content. This process, despite the fact that it is fatal for the stinging individual, is an involuntary altruistic act advantageous to the colony as a whole by thorough and quick envenomization of the opponent. Selectively, this is perhaps the reason why such a defensive method develops. Sting autothysis in *Pogonomyrmex* is supposedly intended primarily for vertebrates, especially mammals. Among mammals, the soil-burrowing and seed-eating rodents are important, since the grain stores of seed-collecting ants are highly attractive to them (Hermann, 1971). Correspondingly, the toxicity of these ants' venom is extremely high for vertebrates, and only moderate for insects (Schmidt and Blum, 1978).

A very peculiar application of proteinaceous posion gland secretions for defensive purposes was recently uncovered in some Southeast Asian ponerines of the genus *Pachycondyla* (Maschwitz *et al.*, in press). These ants are able to sting humans very painfully and with effects lasting for several hours. However, they can also make use of their poison without injecting it into the opponent. During an attack by other ants, or if kept with a forceps, they spray a fine-bubbled foam in the form of thin files more than 10 cm long, or as foam lumps, consisting of lathered poison secretion (Figure 4.6). This is achieved by the release of droplets of poison from the tip of the sting within the sting chamber near the cloacal opening. The droplets then are lathered with air, which supposedly comes from the spiracular plate stigmata. The formation of foam does not result from a special compound of the secretion but is due to the general ability of proteins to form froth (e.g., whipped white of eggs). The Dufour's gland, the lipophilous contents of which would diminish the foaming capacity of the poison, however, is nearly fully reduced here. Attacking ants that

FIGURE 4.6. *Pachycondyla tridentata* worker, when attacked by *Pheidole* sp. workers, ejects its poison gland secretion as a foam toward the aggressors.

come in contact with the secretion or that are pasted over with the foam immediately cease aggression because of the mechanical impediment, and begin to clean themselves.

Apart from proteins the poison glands of stinging ants can also contain volatile trail and alarm pheromones and also sexual pheromones (Buschinger, 1972, 1975, 1976; Hölldobler, 1971). However, none of these substances has yet been identified chemically.

Among the ants capable of stinging, the myrmicine genus *Solenopsis* represents an exception. In addition to a small amount of proteins, a high number of pharmacologically very active 2,6-dialkylpiperidines are produced in their poison glands (Blum and Hermann, 1978a).

The Dufour's gland, which also opens into the aculeus in ants with a functional injection stinger, is a simple glandular tube in which the secretions are accumulated. It contains saturated or unsaturated aliphatic hydrocarbons in the Myrmeciinae, Ponerinae, and Myrmicinae that have been investigated (see reviews in Blum and Hermann,

1978a; Parry and Morgan, 1979). The primary functions of the Dufour's gland secretions are not yet elucidated; perhaps they serve in lubrication of the movable parts of the aculeus. Additional functions such as trail or alarm substances have been reported only in myrmicines (Blum and Hermann, 1978a; Hölldobler, 1977).

Modification and Reduction of the Pugnatory Stinger

Despite the fact that the hydrophilous poison secretions of Formicidae are potent venoms, the majority of ants have abandoned the sting apparatus as an injection weapon and fight by other means. This phenomenon, known as sting reduction, has been the subject of many myrmecological studies (Forel, 1878; Stumper, 1960; Maschwitz and Kloft, 1971; Maschwitz, 1975; Kugler, 1979a). Two of the most species-rich subfamilies, the Formicinae and Dolichoderinae, are directly characterized by their lack of a functional stinging apparatus. Also, many myrmicines and the old world army ants (with the exception of *Aenictus*) no longer possess functional injection stinging apparatus.

The reduction process has evolved polyphyletically. In different ant groups it is variably pronounced and advanced to variable degrees. Among the Myrmicinae, all possible intergradations, from weakly to rather extensively reduced organs, can be seen (Kugler, 1979a).

We do not present here in detail the steps of reduction as described by Kugler (1979a), the diminution of single parts or the whole stinging apparatus with its cuticular elements, the modification or disappearing of single sting elements, and so on. Instead, we discuss the basic aspects of functional morphology.

In principle, two evolutionary directions can be observed (Maschwitz, 1975): First, the stinging apparatus was replaced by other chemical or mechanical fighting mechanisms, has been abandoned completely as a fighting organ, is reduced together with the glands, and has undergone a functional change. This applies to the Dolichoderinae, which fight with their Janet's gland (see below) and whose glandular stinging apparatus has become very tiny (Figure 4.7). It can be expected that this tiny relic organ nevertheless has an unknown function, since organs without function, the development of which costs energy and material, in a strictly evolutionary reflection are not conceivable (Gutmann and Bonik, 1981). The reduction process is more comprehensible in the mandible-fighting Attini, where the small stinging apparatus, which is not capable of stinging, produces trail pheromones in its poison gland instead of defensive secretions.

In *Atta* species, the trail pheromones have been identified as methyl-4-methylpyrrole-2-carboxylate (Tumlinson *et al.*, 1971) and as 3-ethyl-2,5-dimethylpyrazine (Cross *et al.*, 1979).

Second, the stinging apparatus remains a fighting organ; however, instead of the original aqueous protein venoms, lipophilic poisonous and repellent substances are produced that can be applied to the surface of the opponent and need not be injected. At first, the penetration function of the sting is lost, the aculeus and the valves are enlarged at the same time (Kugler, 1979a) for quicker release of the surface venoms. Finally, the gland contents are released through muscle and hemolymph pressure, and the pumping structures atrophy. The best known instance of such a development is the highly modified poison gland of the Formicinae, the formic acid secretion of which is tipped or even sprayed at the opponent (e.g., *Formica* or *Polyrhachis*, Figure 4.7). Additionally, the lipophilic secretions of the Dufour's gland are ejected. Aside from esters, ketones, and alcohols they consist mainly of hydrocarbons that facilitate as carriers the penetration of the formic acid through the lipophilic arthropod cuticle (see summary in Blum and Hermann, 1978a). The pumping and injecting apparatus is strongly modified and reduced, and serves as a supporting element of the venom duct.

Considerably less reduced is the stinging apparatus in *Monomorium pharaonis*. In this species, a secretion that contains several alkaloids (Ritter *et al.*, 1975, monomorine 1, 2 and 3) is discharged out of the poison gland by means of the typical pumping mechanism of the stinging apparatus. But it is superficially applied to the opponent with the tip of the stinger. The secretion has a strong insecticide effect (Levinson *et al.*, 1974). In addition, it is an effective repellent (Hölldobler, 1973) and acts as trail pheromone (Ritter *et al.*, 1975).

The presence of limonene and other terpenes in lipophilic and insecticide poison gland secretions have been found in *Myrmicaria* species with stinging apparatuses incapable of stinging (Quilico *et al.*, 1961; Brand *et al.*, 1974; Howse, 1975). According to Howse (1975), *Myrmicaria eumenoides* is several orders of magnitude less sensitive to its own secretion than are other insects, including ants.

In *Pheidole fallax*, the fighting soldier subcaste, but not the ordinary workes, produces a strongly rank-smelling lipophilic indole, likely skatole, in the hypertrophied poison gland, which has a repellent function (Law *et al.*, 1965). The Dufour's gland is largely reduced. The *Pheidole* species possess, as far as is known, strongly reduced stinging apparatuses incapable of stinging (Foerster, 1912; Kugler, 1979a). Instead of the poison gland, the Dufour's gland can also take over the main task for fighting. This is known from

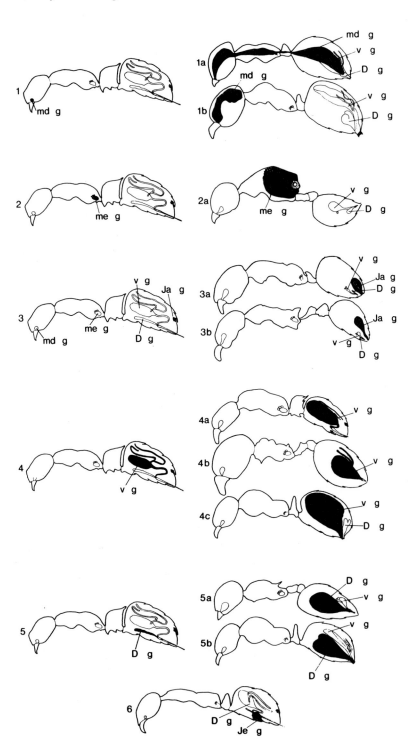

Cremastogaster scutellaris (Maschwitz, 1975) in which the volume of Dufour's gland exceeds by several times that of the poison gland (Figure 4.7). The lipophilic secretion during fighting is emitted not only at the tip but also at the base of the reduced aculeus, which is no more suitable for stinging. It is smeared onto the opponent with the aid of a very mobile gaster that is raised over the alitrunk of the ant. An apparently analogous situation is represented by the myrmicine, *Zacryptocerus multi-spinosus biguttatus*, which also has a reduced injection apparatus and a greatly enlarged Dufour's gland, and which releases a lipophilic secretion in the same body posture as *Cremastogaster* (Coyle, 1966; Kugler, 1978). Hypertrophied Dufour's glands occur also in some dulotic *Formica* species (Regnier and Wilson, 1971). Further investigations of myrmicines with reduced injection sting elements will presumably reveal additional interesting variations with respect to poison and Dufour's gland physiology.

FIGURE 4.7. Glandular organs used for the secretion of defensive substances in ants. (*Left*) Ordinary glandular equipment of a ponerine ant (*Diacamma*). (*Drawn in black*) The organ, in normal size, which has become an oversized or specialized fighting organ in the species on the right side. (1) Mandibular gland (md g). (a) In *Camponotus* (*Colobopsis*) sp. near *saundersi*, where the secretion can be released by bursting of the worker (Maschwitz and Maschwitz, 1974). (b) In *Lasius fuliginosus* (Pavan, 1956). (2) Metapleural gland (me g). (a) In *Cremastogaster inflata* the extremely enlarged gland produces a sticky defensive secretion (Maschwitz, 1974). (3) Janet's gland (Ja g), tergal intersegmental gland between abdominal segments 6 and 7. (a) With defensive function in Dolichoderinae, e.g., *Iridomyrmex humilis* (Pavan and Ronchetti, 1955, "anal gland"); (b) with defensive function in *Pheidole biconstricta* (Kugler, 1979b). (4) Venom gland of the sting apparatus (v g). (a) In *Pachycondyla tridentata*. The proteinaceous secretion of the hypertrophied gland is applied to the opponent as a foam (Maschwitz *et al.*, in press). Dufour's gland is atrophied. (b) In *Pheidole fallax* soldier. The repellent secretion, containing an indole, is applied to the surface of the opponent (Law *et al.*, 1965). Dufour's gland is atrophied. (c) In Formicinae, e.g., *Formica polyctena*, where the highly modified venom gland produces formic acid. (5) Dufour's gland of the sting apparatus (D g). (a) In *Cremastogaster scutellaris*. The lipophilic secretion of the hypertrophied gland is applied to the surface of the opponent (Maschwitz, 1975). (b) In *Formica subintegra*. The hypertrophied gland produces a "propaganda pheromone" used during slave raids (Wilson and Regnier, 1971). (6) Jessen's gland (Je g). In *Leptogenys ocellifera* this sternal intersegmental gland between abdominal segments 5 and 6 produces a defensive secretion (Jessen *et al.*, 1979). In *Diacamma* and other ants this gland is lacking.

Defensive Glands and Glandular Secretions

Apart from the glands of the stinging apparatus a number of other complex glands are present in ants, the secretions of which may be used as chemical weapons: the mandibular glands, metapleural glands, and two gastral glands ("Janet's gland," also known as tergal, pygidial, or anal gland, and "Jessen's gland" (Figure 4.7). In a yet unidentified Mediterranean *Messor* species the rank-smelling feces are used for defense (Maschwitz, 1975).

Mandibular Glands

Mandibular glands have been found in the Hymenoptera as well as in most other insects. They are present in all ants. The mandibular glands are a pair of saclike complex glands that open at the base of the mandibles. They are found not only in workers but also in queens and males; however, the composition of the secretions may vary between the castes. It can be assumed that the mandibular gland has a basic function, though this is not yet known. Its secretions ordinarily consist of a blend of lipophilic, strongly smelling substances; in workers, ketones, aldehydes, or esters are predominant. Aliphatic ketones have often been found, as, for example, 4-methyl-3-heptanone in the Ponerinae, Dorylinae, Pseudomyrmecinae, and Myrmicinae. Sulfurous compounds such as the strongly foul-smelling dimethyldi- and trisulfides are known, as well as nitrogen-containing compounds (for a review covering morphology and chemistry, see Blum and Hermann, 1978a,b).

Little considered in the context of defensive behavior has been the fact that males also have strongly smelling mandibular gland secretions. It is assumed that they are released not only for sexual reasons, but also in cases of peril. However, males by no means engage in colony defense (Blum and Hermann, 1978a). For all the substances mentioned above a repellent action and partly also an insect toxicity may be expected. They also can act as combat secretions when released by the workers during biting. The combat function of mandibular gland secretion should be rather common, though it has been scarcely investigated (e.g., Ghent, 1961; Regnier and Wilson, 1969; Bergström and Löfqvist, 1971; Leuthold and Schlunegger, 1973; Blum et al., 1969; Brough, 1978).

Mandibular gland secretions in ants, as far as they have a combat function, are always supplementary weapons. In all instances, further fighting mechanisms are present, mostly chemical ones. Frequently, the mandibular gland secretions have an additional function that is important for fighting; as alarm pheromones they serve to coordinate the activities of the colony members (see Alarm Behavior and Organization of Social Defense).

Only a few instances are known in which the mandibular glands might be specialized exclusively as combat organs. This is the case in *Lasius fuliginosus* (Pavan, 1956; Bernardi *et al.*, 1967; Maschwitz and Kloft, 1971), the mandibular glands of which are highly enlarged and occupy a major part of the head capsule volume (Figure 4.7). One of the compounds of their secretion, dendrolasin, has a strong insect toxicity which is said to reach, especially for ants, the power of DDT.

A peculiar transformation of the typical mandibular gland has been found in a few Southeast Asian *Camponotus* species of the subgenus *Colobopsis* (Maschwitz and Maschwitz, 1974; Figures 4.7 and 4.9). In *C. saundersi* and another closely related species the glandular organs in workers are extremely hypertrophied and occupy not only the major part of the head capsule but reach with thick tubes through the body until the end of the gaster. They contain a strongly sticking secretion that the ants smear on their opponents during fighting. But the secretion is not only discharged through the glandular opening at the base of the mandibles. During violent fighting the substances may also be released through strong contractions of the worker's gaster until it bursts along the intersegmental membranes. The attackers become stuck by the secretion of the simultaneously bursting glandular tubes. In this event, an example of extreme altruism, the workers sacrifice themselves for the welfare of the society. Hypertrophied mandibular glands with a defensive secretion are also found in *Acanthomyops claviger* (Ghent, 1961; Chadha *et al.*, 1962) and *Calomyrmex* sp. (Brough, 1978; Brown and Moore, 1979).

Metapleural Glands

The metapleural glands, which at present are known only in ants, are situated at the posterolateral corners of the alitrunk. Ordinarily they consist of a cuticular chamber or atrium that opens to the outside with a pore or a slit. A number of bundled single gland cells with separate ducts through the cranial wall open into this reservoir (review Pavan and Ronchetti, 1955; Maschwitz, 1974; Blum and Hermann, 1978a). In the Formicinae, the cuticular chamber is transformed to an open groove. The organ is lacking only in some social parasitic ants, and in some males (Gösswald, 1953; Brown,

1968). It was suggested for a long time that this gland would produce pheromones for recognition and identification of nest mates and alien species (Brown, 1968).

First investigations of the secretion in *Atta* (Maschwitz *et al.*, 1970) revealed that it was an acid, a highly antibiotically active substance that contained phenylacetic acid. Since no pheromone function could be observed, Maschwitz *et al.* (1970) interpreted the organ as an antiseptic weapon against microorganisms in the ant colony. Further investigations of Schildknecht and Koob (1971) in *Atta* and other myrmicines revealed the presence of fungistatic β-hydroxy fatty acids. They also detected indole-3-acetic acid, which is, in higher concentrations, a plant auxin. Whether it also interferes with the growth of fungal microorganisms has not yet been elucidated.

A comparative study by Maschwitz (1974) in ants of the subfamilies Myrmicinae, Myrmeciinae, Pseudomyrmecinae, Ponerinae, Aneuretinae, and Dolichoderinae revealed the presence of antibiotically active, acid glandular secretions in all these groups. The secretions flow out of the reservoir continuously, since it cannot be locked, onto the body surface and the nest material.

In some myrmicines of the subgenus *Physocrema*, genus *Cremastogaster*, the metapleural glands are extremely hypertrophied, for example, in *C. inflata*, in which a sticky, ropy secretion is produced (Figures 4.7 and 4.8). When irritated or seized the workers extrude a droplet of the secretion, which can be retracted again. During fights the workers smear the opponents with the substance and thus impede them seriously in their ability to move. The metapleural gland in this case has gone through a change of function and has become a combat organ for larger opponents, since in its secretion antibiotically active compounds can no longer be identified.

Another *Physocrema* species, *C. difformis*, in its hypertrophied glands, produces instead of a sticky secretion a substance with a strong phenolic smell that supposedly acts as a venom and repellent.

Janet's Gland

Janet's gland is also known as the "anal gland" (Forel, 1878), "tergal gland" (Hölldobler and Engel, 1978; Jessen *et al.*, 1979), and "pygidial gland" (Kugler, 1979b).

Janet's gland opens through the intersegmental membrane between the sixth and seventh tergum. The latter, together with sternum VII, forms the abdominal end of the ant's body. Thus, the

FIGURE 4.8. *Cremastogaster inflata* when seized with a forceps releases droplets of sticky metapleural gland secretion from the glandular orifices.

term "tergal gland," which was proposed for this gland, is as inaccurate as the term "pygidial gland." The first name could indicate as much as three different glands situated under the gastric terga, and in the second case the glands could open between the sixth and seventh or between the seventh and eighth terga. Furthermore, the term "pygidium" has been applied to the last abdominal segment of insects, without a coelom. "Anal gland" is not correct, since the gland does not open into the rectum (Pavan and Ronchetti, 1955). Therefore, we propose to name it after the author who first localized the organ exactly, Janet (1898).

In Dolichoderinae, this gland has been known as a combat gland for a long time (Forel, 1878). It consists of a bilobate reservoir with a common opening. At both sides of the reservoir lobes a number of unicellular glands are situated, the channels of which join to form a common duct, or lead separately into the reservoir (Maschwitz and Kloft, 1971; Blum and Hermann, 1978b). A homologous organ is also found in most other subfamilies, the Myrmeciinae, Nothomyrmeciinae, Myrmicinae, Pseudomyrmecinae, Dorylinae, and Aneuretinae. It seems to be lacking only in Formicinae (Hölldobler and

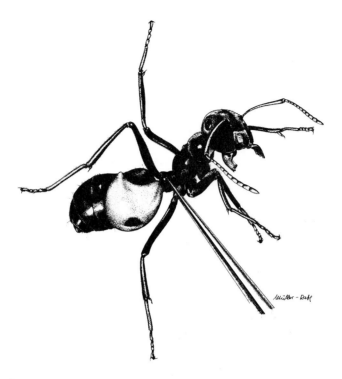

FIGURE 4.9. *Camponotus* (*Colobopsis*) sp. near *saundersi* when seized with a forceps bursts and releases the secretion of the hypertrophied mandibular glands.

Engel, 1978). In these subfamiles this organ is generally less differentiated and smaller than in the Dolichoderinae, and it consists of a paired cluster of glandular cells that open either into two reservoirs, into a bilobate-fused reservoir, or without a reservoir directly into the intersegmental membrane fold between segments VI and VII.

In a number of dolichoderines the composition of Janet's glands secretions is well investigated (for a summary see Blum and Hermann, 1978b). Apart from strongly smelling short-chained fatty acids, alcohols, and ketones, cyclopentanoid monoterpenes are often found, for example, iridodial or iridomyrmecin, the latter possessing strong insecticide effects resembling DDT in toxicity. The Dolichoderinae smear their opponents with these substances during fights. In addition to their combat function the secretions also have alarm effects. This also applies to the Janet's gland of *Pheidole biconstricta* (Kugler, 1979b). In *Novomessor*, this gland also appears to have a defensive function (Hölldobler *et al.*, 1976). In some Ponerinae the secretions of workers act as trail and recruitment pheromones (Maschwitz and Schönegge, 1977; Hölldobler and Engel, 1978). The basic function of Janet's gland is, however, not yet known.

Jessen's Gland

While investigating the ponerine, *Leptogenys ocellifera* (Maschwitz and Mühlenberg, 1975) we realized that this quickly and painfully stinging ant produced a distasteful tarlike smell when it was touched or seized, a typical indication of the existence of an additional chemical weapon. The search for the source of the odorous substance revealed that it was not produced in one of the ordinary complex glands of ants, but in a previously unknown large complex gland between abdominal segments V and VI (Figure 4.7). It was more closely investigated by Jessen (1977). Since there exist other sternal glands in ants (e.g., "Pavan's gland," Pavan and Ronchetti, 1955; Jessen *et al.*, 1979; Hölldobler and Wilson, 1978; Hölldobler and Engel, 1978), we propose to name this gland after the author in order to have an unequivocal indication. The gland consists of a large, unpaired reservoir that is locked by a lamellum. Two types of single gland cells open separately into this reservoir. In addition, a glandular palisade epithelium lies on the sixth sternite, near the opening of the reservoir. As with some Janet's glands, Jessen's gland is also associated with peculiar cuticular squamous structures near the orifice (Jessen *et al.*, 1979). The gland has been found in further *Leptogenys* species as well, but it is lacking in most other ponerines investigated.

ALARM BEHAVIOR AND ORGANIZATION OF SOCIAL DEFENSE

Group defense, joint defensive actions of several specimens, are quite common in animals (Edmunds, 1974). In social insects, organized group defense is one of the most conspicuous behaviors, and one of the most important foundations of their ecological success. We present a survey primarily of behavioral mechanisms involved in the organization of defense in ants. Communication, with regard to alarm behavior, is a basic requirement for group defense.

Alarm Behavior of Ants

Little is known about alarm mechanisms in ants that work without chemical cues. Thus, Markl (1967) investigated the stridulatory signals of leaf-cutting ants (*Atta* and *Acromyrmex*) buried under soil. Other ants of the colony are attracted by this sound and dig out the stridulating nest mate if it is not covered by more than 3 cm of soil. Similar observations were reported by Collart (1925) for *Megaponera foetens* and by McGurk *et al.* (1966) for *Pogonomyrmex barbatus.* In *Novomessor*, substrate-borne stridulatory signals are said to play a role in prey retrieval (Markl and Hölldobler, 1978), and in *Leptogenys chinensis*, chirping, together with chemical signals, is involved in the organization of colony emigration (Maschwitz and Schönegge, 1977). However, stridulation is a very common behavior in numerous ant species, as in many solitary insects, and its meaning in most cases is not yet fully understood. Sudd (1967) claims that stridulatory sounds in general are intended for predators, and that vibration might be as potent a warning to an arthropod predator as aposematic coloration to a vertebrate.

Carpenter ants of the genus *Camponotus* exhibit another kind of vibrational alarm behavior when disturbed. They rap on the substrate with the aid of the mandibles or the gaster. The reactions of nest mates who perceive the signal as substrate vibrations depend on the ant's excitation level. They either freeze motionless when little excited, or if more excited before the rapping signal, they increase their running speed and have a lowered threshold of aggression toward moving objects. The authors claim that rapping thus fulfills functions of prey–attack and danger–alarm systems in that it amplifies the effect of other attack-releasing stimuli (Markl and Fuchs, 1972).

Hölldobler (1977) records a vibrational alarm communication behavior in an Australian ponerine, *Amblyopone australis*, that is apparently transmitted entirely by tactile contacts between nest mates. In *Oecophylla longinoda*, Bradshaw *et al.* (1975) observed that major workers strike their bodies against the leaf nest when it is disturbed, making a sound similar to dropping of peas onto a plate. Finally, workers of *Lasius, Messor*, and *Tapinoma* species often exhibit a peculiar jerking movement when the nest is opened. In other species the excited ants simultaneously hit the nest mates with the antennae or even with the head. Goetsch (1953) interprets this behavior as an alarm signal that stimulates the carrying of larvae into sheltered places of the nest. Tactile signals also play a role in the invitational behavior during alarm recruitment with chemical trails (see following).

Apart from these few examples of vibrational or tactile alarm signals, the most important cues of alarm, and thus also defensive communication, are various chemical secretions, that is, pheromones.

Recent reviews on this subject were given by Gabba and Pavan (1970), Hölldobler (1977, 1978), and Parry and Morgan (1979).

First of all it must be stressed that a chemical alarm defense system has been found in most ant species investigated (reviews by Parry and Morgan, 1979; Blum and Hermann, 1978a), even in those with a vibrational alarm signal mentioned above. For example, Vowles (1952) demonstrated that formic acid elicits attack in larger *Camponotus* workers, and a flight response in the smallest ones. The chemical alarm system in other formicine species was investigated by Maschwitz (1964), Regnier and Wilson (1968, 1969) and Wilson and Regnier (1971), and in leaf-cutting Attini by Riley *et al.* (1974a,b) and Crewe and Blum (1972). Bradshaw *et al.* (1975) studied the alarm pheromones of the weaver ant, *Oecophylla.*

Sources of alarm pheromones are various exocrine glands, and often primarily defensive secretions, such as formic acid, simultaneously function as an alarm substance (Löfqvist, 1976). Following Hölldobler (1978), the mandibular gland produces alarm pheromones in Ponerinae, Myrmicinae, and Formicinae. In Myrmicinae and Formicinae the poison and/or Dufour's gland secretions may function as alarm substances; they also may serve as trail pheromones.

In the myrmicine, *Zacryptocerus varians*, 4-heptanone was identified as an alarm pheromone from the mandibular glands (Olubajo *et al.* 1980), and methyl anthranilate was found in the mandibular gland secretions of *Aphaenogaster fulva* and *Xenomyrmex floridanus*. It initiates escape behavior in these species (Duffield *et al.*, 1980).

In Dolichoderinae, Janet's gland is the source of both an alarm pheromone and a defensive secretion. A recently discovered Janet's gland in a myrmicine, *Pheidole biconstricta*, also serves both these functions (Kugler, 1979b), and in the myrmicine, *Cremastogaster inflata*, the highly hypertrophied metapleural glands contain a sticky defensive secretion that also releases alarm behavior (Maschwitz, 1974). In *Cremastogaster scutellaris*, a general frenzy alarm in aggregations of workers inside and outside the nest is mediated by the defensive secretions from the stinging apparatus, whereas the mandibular gland secretion on trails outside the nest releases an alarm behavior with a concentration of workers around the alarm center (Leuthold and Schlunegger, 1973). Finally, Dufour's glands in Formicinae produce an alarm pheromone. Thus, in *Formica*, the alarm communication comprises pheromones from three different glands, the mandibular, poison, and Dufour's gland (Maschwitz, 1964). Such alarm communication systems involving the secretions of several glands in one species are quite common in ants. On the other hand, it is as well a widespread phenomenon that the multicompound secretions of one gland serve several functions (Howse *et al.*, 1977; Brough, 1978).

In many ant species the discharge of alarm and defensive secretions is indicated by characteristic movements or body postures. *Formica* workers bend the gasters forward beneath their legs and spray a mixture of formic acid and Dufour's gland secretion. *Cremastogaster* (Myrmicinae) raise their gasters to a vertical position while releasing a defensive secretion through the sting and alarm pheromones from the mandibular glands (Blum *et al.*, 1969; Leuthold and Schlunegger, 1973; Hölldobler, 1978). *Zacryptocerus umbraculatus* and *Z. multispinosus biguttatus*, like *Cremastogaster*, have special-shaped, flattened petioles that facilitate the raising of the gasters for release of the defensive secretion, whereas *Z. multispinus* has a normal-shaped petiole, does not emit a defensive substance from the apex of the gaster, and is protected by quick escape movements and a well-developed cuticular armament (Coyle, 1966). Similar gaster raising was observed in other myrmicines (*Monomorium pharaonis, Diplorhoptrum fugax*), in dolichoderines, and in formicines (e.g., *Calomyrmex*, Brough, 1978; *Oecophylla*, Hölldobler and Wilson, 1977a).

From the behavioral point of view, the reactions of nest mates to various alarm substances are as interesting as the effectiveness of the alarm system itself.

As was shown by Wilson and Bossert (1963), the physical properties of the volatile alarm substances seem well designed for an economical alarm system. The alarm pheromone of the harvester ant, *Pogonomyrmex badius*, is produced in the mandibular gland. The amount of pheromone in the glandular reservoirs of one worker is sufficient to provide only a brief signal, which expands in still air to a maximum radius of about 6 cm in 13 sec, and fades out in 35 sec. In case of a persistent danger, the ants previously alerted by the alarm pheromone of a nest mate will discharge their own pheromone. The alarm thus increases and may spread through the entire colony (Hölldobler, 1978). Low concentrations of the alarm substance at the periphery of a pheromone cloud release only attraction behavior, whereas higher concentrations provoke real alarm and aggression or, sometimes, flight (Maschwitz, 1964). Further examples are given by Hölldobler (1978).

In other ant species a differentiated sequence of responses is released by multicomponent alarm substances. An impressive example is described by Bradshaw *et al.* (1975) in *Oecophylla longinoda*. Their mandibular gland secretion contains at least 33 chemicals. The authors tested the reactions of ants on several of the major components, and they found that the most volatile compound, hexanal, with the widest "active space," released alerting and alarm signals with little evidence of directional orientation. The less volatile 1-hexanol attracts the ants from a distance of about 10 cm, but stops or repels them a few millimeters from the source. Other slightly volatile compounds control biting behavior. Thus, the activities of the alerted ants are successively centered toward the site of release of the alarm substance, where aggressive behavior is elicited.

The secretions of the sting glands sometimes contain attractive compounds and often they are used as trail substances. In such cases it may become difficult to discriminate between a trail pheromone and alarm function of the secretion. Thus, some ant species, for example, *Myrmica rubra* or *Pheidole dentata*, organize colony defense by recruitment of nest mates to a source of disturbance, mainly caused by ants of other species within the territory of the colony (Cammaerts-Tricot, 1974; Wilson, 1976). *Myrmica* workers deposit attractive Dufour's gland secretions around an ant of another species, then lay a trail of poison gland secretion to the nest and finally walk back to the intruder, depositing a trail of attractive Dufour's gland secretion. A number of nest mates follow this trail to the site where help is needed. The trail recruitment system of *Pheidole dentata* (Wilson, 1976) seems mainly important for defense of the colony against invading *Solenopsis* species. Recruitment is achieved by a trail substance from the poison gland. This behavior is evoked most strongly by fire ants and other *Solenopsis* species, but less by other ant species, and not at all by attacks, for example, of vertebrates disturbing the nest.

European *Pheidole pallidula*, however, do not differentiate between various other ant species. They recruit numerous nest mates for defense against all other sympatric species, the workers of which are placed into the *Pheidole* territory (e.g., *Diplorhoptram banyulensis*, *Cremastogaster scutellaris*, and several Formicinae species; Stumm and Maschwitz, unpublished). *Technomyrmex albipes* recruits masses of workers to localities where their territory contacts that of an *Oecophylla smaragdina* colony, for example, on a neighboring tree (U. Maschwitz, unpublished). When the attacking *Technomyrmex* have killed a considerable number of the opponents with the aid of their Janet's gland secretion, the *Oecophylla* react with a well-organized escape of the entire colony (Figure 4.3). *Oecophylla longinoda* has two different recruitment systems in reactions against intruders, a short-range system that involves a trail from a sternal gland and a long-range recruitment, mediated by odor trails from a rectal gland and by tactile stimuli (Hölldobler and Wilson, 1978). Recruitment by means of a hindgut pheromone trail also occurs during intraspecific territorial encounters of *Myrmecocystus* (Hölldobler, 1976b; see below). These observations indicate that alarm-recruitment systems might be quite common in ants.

Defense of Food Resources

Defense of the colony, or the nest, is a crucial element in the protective system of ants. However, it is also important for many ant species to protect their food resources. In this section we discuss a number of mechanisms that serve to defend food resources against predators and other animals competing for food. Intraspecific competition is treated in the following section.

Arthropod prey and honeydew from aphids, scale insects, or membracid cicadas are the prevailing constituents in the diet of most ants. Free living arthropods often stroll around in a rather unpredictable way, and thus there is little opportunity to protect them from being captured by other predators. Ants that have caught prey of moderate size usually carry it, alone or with the help of nest mates, quickly to the nest. However, if ants detect prey or a food source too large for removing it *in toto*, they often take measures to protect it from competitors.

Wilson (1971) distinguished three types of behavior in ants who have found a piece of food, for example, a sugar bait. "Opportunists" like *Paratrechina longicornis* are swift running, quickly find the food, fill their crops, and recruit nest mates with an odor trail. But they are timid and draw back when other ant species arrive. The "extirpators," including *Pheidole* species with soldiers, or *Solenopsis geminata*, recruit nest mates to the food and fight against competitors that are already present. The third group, the "insinuators," for example, *Tetramorium simillimum* or *Cardiocondyla* species, rely on small size and stealthy behavior to gain the food. They sneak through the crowds of "extirpators" around the bait and reach it without eliciting the aggressive behavior of competitors.

In addition to physical fighting, chemical strategies may be involved in the defense of a food source. Thus, the Pharaoh's ant (*Monomorium pharaonis*) deposits poison gland secretion on large prey objects. This secretion has a strong repellent effect for other ant species. Similarly, *Monomorium minimum* expels other ants, for example, *Solenopsis invicta*, with the aid of its powerful poison from food sources (Baroni-Urbani and Kannowski, 1974). *Diplorhoptrum fugax*, a small myrmicine-like *Monomorium*, usually preys on the brood of foreign ant species. This ant lays odorous trails in narrow tunnels leading to brood chambers of adjacent ant nests. There again a poison gland secretion is discharged on the brood. The long-lasting repellent effect prevents the foreign ants from defending their own immatures against the predators (Hölldobler, 1973).

More intricate relations occur between ants and plant lice, the source of honeydew. These insects, which often settle in large colonies, remain for weeks or months in the same localities on bark, twigs, leaves, or stems of herbs. The correlations between ants and plant lice vary in a wide range, from those visited only occasionally by ants to species completely dependent on their persistent care, and which have developed conspicuous adaptations for this symbiosis (for aphids, see Kunkel, 1973, where further literature is cited).

Ants of the genus *Formica* are often observed guarding plant lice colonies and defending them against predators or parasites (Wheeler, 1910; Sudd, 1967; Wilson, 1971). Other ant species construct tents, or pavillions for their "cattle" herds, namely, *Cremastogaster, Oecophylla*, and *Lasius* species (Wheeler, 1910). The scale insect *Hippeococcus* is found in Java living with *Hypoclinea* (Dolichoderinae) species. When the nest is disturbed, the ants, using their antennae, touch the *Hippeococcus*, which then climb on the ant's backs and are carried away (Reyne, 1954). Some ant species have aphid eggs in their nests during winter, and they lick the eggs and care for them as if they were their own (Pontin, 1961). In spring, these eggs or the newly hatched plant lice are sometimes carried to stems and roots of plants, for example, by *Lasius americanus* (Wheeler, 1910), with root aphids (*Aphis maidiradicis*) in corn fields. Many ant species attend root aphids directly in their nests, for example, *Tetramorium* or *Lasius flavus*, where their "cattle" are obviously best protected.

Mutualism between certain plants and ant species defending them is reported from numerous associations, most of which occur in the tropics. Thus, the swollen-thorn acacias in the New World provide both food and domicile to the ant, *Pseudomyrmex ferruginea*. The ant protects the acacia from phytophagous insects as well as from neighboring plants, especially climbing vines (Janzen, 1966, 1967). In a Nigerian rain forest, a similar relationship occurs between the pseudomyrmecine, *Pachysima*, and trees of the genus *Barteria*. The ants chew off the foliage of plants below the tree or growing into the *Barteria* crown from the side. They also clean the leaves from epiphyllae and debris. On the other hand, *Pachysima* gains its food mainly from the Homoptera tended inside the hollow branches of the tree (Janzen, 1972).

Defense of a Territory

As already mentioned, some ant species display rather specific alarm defense behaviors against intruders or aggressors belonging to alien ant species. However, an ant colony must also defend its territory, nest, and foraging area against competing colonies of its own species. Territoriality in social insects was recently revised by Baroni-Urbani (1979). Hence, we restrict our discussion to the behavioral mechanisms involved in a few well-documented examples.

A territory may be defined as an area that an animal, or a colony, uses exclusively and that it defends against intruders, conspecifics, and sometimes also other species (Hölldobler, 1979). It is well known that ants of the same species often partition their activities in the environment; however, their territorial strategies have been investigated only recently, and in only a few species.

Several authors have stressed that territories in most ants are by no means sharply defined. Often there is a considerable spatial overlap (Baroni-Urbani, 1979; Pickles, 1935, 1936; Talbot, 1943). On the contrary, Hölldobler (1976a) found that harvester ants of the genus *Pogonomyrmex* establish a system of long-lasting trunk trails around their nests, which do not overlap. The foraging grounds thus are subtly partitioned and aggressive encounters only occur when workers of two colonies are forced to forage at the same food source.

Carrying of alien conspecifics out of their own territory was described by Dobrzański (1966) in *Leptothorax acervorum*. Similarly, *Acromyrmex versicolor* seems to eject noncolony members out of the territory by carrying (Gamboa, 1975), and *Manica rubida* removes *Myrmica scabrinodis* workers from the nest surface and releases them uninjured a few decimeters away (Le Masne, 1965). However, the territorial fights, especially between conspecific ants, are often extremely fierce. Zakharov (1969), De Bruyn and Mabelis (1972), and Breen (1976) reported so-called "spring-battles" of *Formica polyctena* and *F. lugubris* that sometimes ended with the destruction of one colony by the other. Forel (1948) describes fierce battles between *Tetramorium* colonies that lasted for more than one month. *Pseudomyrmex ferruginea*, which inhabit swollen thorn acacias, attack most animals on their trees; however, they become most aggressive toward conspecifics when one colony expands its territory. Worker mortality in these battles is extremely high (Janzen, 1967).

With respect to the differences in aggressiveness during intra- and interspecific encounters, the data of various authors are somewhat contradictory. According to Wilson (1971), Carrol and Janzen (1973), and Hölldobler and Wilson (1977a), hostile behavior should be strongest between competing societies of the same species, and it should decrease with increasing taxonomic differences between the interacting species (Gause's Law). However, De Vroey (1979) found that in *Myrmica rubra* the aggression is lower toward conspecifics of alien nests than toward other sympatric ant species. This might be due to the polygynous colony organization of *M. rubra* (see below).

A reduction of detrimental fighting between members of two conspecific colonies, as in *Pogonomyrmex*, is also achieved by chemical demarcation of the territories. Cammaerts *et al.* (1977) report a well-documented instance of marking a newly explored territory by *Myrmica rubra* workers. They deposit small traces of Dufour's gland secretion on the substrate that have a short-lived recruitment activity for nest mates and a longer-lasting territorial marking effect.

In the African weaver ant, *Oecophylla longinoda*, the territories are marked by droplets from a specialized part of the rectal sac, a rectal gland (Hölldobler and Wilson, 1977a; Hölldobler, 1979). The territorial pheromone is colony specific, and the colonies exclude one another in a very efficient way. A colony-specific territorial pheromone has also been found in leaf-cutting ants of the genus *Atta*. In this case, the pheromone is produced in one of the newly detected sting chamber glands near the sting (Jaffé *et al.*, 1979; Bazire-Bénazet and Zylberberg, 1979).

Somewhat curious behaviors seemingly involved in interspecific competition were reported by Blancheteau (1975). *Messor barbarus* workers carry small stones toward neighboring nests of *Pheidole pallidula* or *Cataglyphis cursor*, and drop them into their nest entrances. Stone dropping into the nest entrances of *Myrmecocystus* species was also observed in the dolichoderine *Conomyrma bicolor* (Möglich and Alpert, 1979). This behavior is supposed to reduce the foraging activities of the *Myrmecocystus* colonies.

Myrmecocystus mimicus, a honey-pot ant, exhibits a highly ritualized behavior during territorial combats with conspecific neighbors. According to Hölldobler (1976b), the foraging grounds of *Myrmecocystus* colonies frequently overlap. When foragers get into the territory of another colony, where they encounter workers of the resident nest, both display a significant stilt walking. They then recruit nest mates to the locality of the first encounter. Sometimes the groups then exercise real tournaments over several days, but rarely engage in physical fighting. Finally, in some instances, the weaker colony is overrun by the stronger opponents. They kill the queens and integrate the workers and broods of the inferior society into their own colony.

It seems such territorial merging reactions are not unusual in ants. They have been observed in *Formica fusca* (Kutter, 1957), *Leptothorax curvispinosus* (Wilson, 1975b), and *Leptothorax acervorum* (A.

Buschinger and Koch, unpublished). The elimination of potentially competing conspecific colonies is also often achieved by simply killing or driving off the young queens that try to establish their nests in the vicinity of an already existing one (Brian, 1965). A more peaceful means of reducing intraspecific competition may be the integration of young queens into existing colonies, which then are, or become, polygynous. We refer to this mechanism in a later section.

EVOLUTIONARY ASPECTS OF DEFENSIVE SYSTEMS IN ANTS

Several evolutionary traits in the development of defensive behaviors or structures have already been indicated in the previous sections. Thus, the morphological evolution of the stinging apparatus has been described. Here we refer to some of the selection mechanisms that might have influenced this development.

From Sting and Mandible Fighting to Chemical Defense

Primitive ants, such as Ponerinae or Myrmeciinae, possess with their stinging apparatus a potent poison weapon. In addition to killing arthropods, they are also able to fend off vertebrates very effectively. Nevertheless, the stinging apparatus in most ants is more or less reduced or modified so that it is no longer suitable for stinging. Among these ants, the most successful genera, as judged by their numbers of species, have to be classed, such as *Camponotus, Pheidole,* or *Cremastogaster*, all with several hundred species.

Reduction of the venom apparatus process has evolved polyphyletically, several times among the Myrmicinae, and separately in the subfamilies Dolichoderinae and Formicinae, which are directly characterized by their inability to sting. The Old World Dorylinae also are no longer capable of stinging. Instead, these ants possess fighting mechanisms that are much less efficient, at least against vertebrates. This phenomenon, which at the first glance is contradictory to evolutionary theories, is one of continuous interest to myrmecologists.

The reasons for sting reduction certainly are complex. Simple generalizing argumentations such as the transition from carnivorous to omnivorous life habits (Stumper, 1960; Cavill and Robertson, 1965; Kugler, 1978) are not conclusive. The sting not only serves in subduing prey but also in defense, and omnivorous ants have to defend themselves just as much as carnivorous species. Thus, the ants of most of the primitive subfamilies (e.g., Nothomyrmeciinae, Myrmeciinae, Pseudomyrmecinae, and Aneuretinae) consume carbohydrates as well as proteins (Wilson *et al.*, 1956; Taylor, 1978). Among the nearly exclusively seed-eating harvester ants there are a number of species (e.g., *Pogonomyrmex* spp., *Brachyponera senaarensis*, Levieux and Diomande, 1978) well capable of stinging. On the other hand, among the army ants that are extremely specialized as carnivores the Old World dorylines have reduced stinging apparatuses.

On principle, it must be stated that the reduction of an organ can only take place when it is rarely or no longer used. This means that in ants the stinging apparatus being reduced is no longer used as a fighting organ, and consequently other weapons have already been introduced or substituted.

The solitary ancestors of ants, like nearly all other nonparasitic aculeate Hymenoptera, certainly used the stinging apparatus primarily as a weapon for capturing their larval food. Correspondingly, the poison secretions are well adapted for preying on insects. The secretions contain paralyzing venoms that block the synapses (reviewed by Rathmayer, 1978). In these aculeates the defense against enemies is therefore only a secondary function of the stinging apparatus. The stings even of large species such as *Bembex* or *Philanthus*, for example, do not cause long-lasting pain in humans.

In ants, on the contrary, which hunt with the aid of their stinging apparatus, a prey specialization similar to that in solitary aculeates is much less common. Only among a number of Ponerinae and a few Myrmicinae is specialization toward certain kinds of prey known, chiefly termites, other ants, woodlice, beetles, collembolans, and so forth (Ponerinae: see review of Wheeler, 1936; Myrmicinae: Brown and Wilson, 1959). As far as is known, such ants, for example, *Leptogenys* species, make use of their stingers for hunting prey. But those ponerines that are not specialized toward distinct prey animals also use their sting apparatuses as a hunting weapon. In this subfamily sting reduction has not yet been established. Recently it has been observed that some species (*Leptogenys, Harpegnathus*) do not kill their prey but merely paralyze it by stinging, which renders possible a short-termed stockpiling (Maschwitz *et al.*, 1979). Presumably this ability is important for the ponerines because a stockpiling and social distribution of food with the aid of the crop (trophallaxis) is lacking completely or is poorly developed. Furthermore, it is important that the Ponerinae, contrary to other primitive ant subfamilies, feed exclusively or predominantly on animal proteins even in the adult stage. Additional discussion on elaboration and reduction of the sting may be found in Chapter 6 of this volume.

The question arises as to whether such a stinging apparatus is also a suitable weapon against enemies. As we have already mentioned, the main enemies of ants, because of their territorial habits and their frequency, are members of their own family, as well as other arthropods. Since these opponents are highly variable in size, shape, and fighting techniques, a specialization such as that directed toward distinct prey objects is hardly possible.

We have studied the suitability of the sting apparatus for intra-family fighting in different Myrmicinae, a subfamily in which sting reduction often occurs. We found that the fighting of *Myrmica* or *Tetramorium caespitum* against their cogeners with hard sclerotized cuticles is a "tiresome" process. For the hydrophilic sting venom to be effective it has to be injected into the body of the opponent. The sting point can only penetrate through places with very thin cuticle. In order to find such a place, *Myrmica* workers, for instance, must first hold the animals fast by their mandibles. The following search for an appropriate sting position can last up to half an hour. The stung opponent usually dies. For an effective defense the opponent, how-ever, only has to be dislodged or damaged. The defender in this case does not have to involve itself in a dangerous "fight to the death." In a large number of experiments (stinging species fighting against each other and against nonstinging species like *Atta, Camponotus, Formica, Lasius*) we found that in fights with rapidly moving or hard sclerotized opponents, the stinging apparatus is unwieldly and in no way an optimal fighting instrument.

Generally speaking, the position of the hymenopteran aculeus at the posterior end of the body, opposite to the main sensory organs, is unfavorable for exact pointing and quick stinging. This becomes evident in a tendency characteristic of all parasitic and predatory hymenopterans: one or two segments in the middle of the body become constricted and articulate, a process by which the abdomen and the stinger gain much in movability. In ants, at least the second abdominal segment is contracted and limbered as a petiole. In myrmicines, pseudomyrmecines, and some dorylines the third ab-dominal segment, in addition, was strongly contracted and thus maneuverable. Such a postpetiole is also more or less indicated in the Ponerinae and Myrmeciinae. The resulting narrow and limbate middle of the body, which makes the stinger maneuverable, however, also creates disadvantages. It can be easily hurt by bites, and the lodging of musculature for quick and strong movements of the gaster becomes more difficult. This could be the reason for the peculiarly slow searching movements of Myrmicinae when stinging.

Thus, it is conceivable that in all ants, even those with functional stinging apparatuses, additional weapons are present that act quickly and at the surface of the opponent. As was shown in the ponerine genus *Pachycondyla*, this effect may be achieved even with the poison gland secretion itself, if it is discharged as a foam that covers the opponent immediately and impedes it mechanically.

The mechanically acting mandibles may be considered a phylo-genetically old complementary weapon that many ants, even those with a functional stinging apparatus, employ in fighting. Whether there really exist species that fight exclusively with their mandibles, as was suggested for some Attini, *Pheidole* (soldier caste), *Harpagoxenus*, or *Messor* species, remains questionable. *Atta*, which certainly fight chiefly with their mandibles, release alarming and strongly smelling mandibular gland secretions during combats (Blum and Hermann, 1978a). The mandible fighting soldiers of *Pheidole pallidula* have considerably enlarged Dufour's glands, and *Harpagoxenus sublaevis* smears its opponents during fights with sting gland secretions (Winter, 1979).

At present, two techniques of mandible fighting have been observed, the technique of "quick bite and retreat," as in *Atta sexdens* or *Harpagoxenus sublaevis*, where the attacking ant suddenly advances, bites, and quickly retreats, and the technique of "clinch fighting," as in *Pheidole pallidula* soldiers who cling firmly to the opponent and bite off its limbs one after the other.

As with biting, the conditions of quick applicability and surface action are also precisely fulfilled by the new chemical weapons. The frequently occurring lipophilic repellent and insecticide compounds can be applied externally and almost instantly by smearing or spraying the opponent. Sticky substances also occur. The relevant glands and their secretions have already been described.

The mandibular gland, with its odorous secretions, serves pre-sumably in most cases only as a complementary combat organ. However, it seems well adapted as a chemical weapon, since it acts in conjunction with the mechanically acting mandibles and often serves as an alarm releaser during fighting. Mandibular glands with the above-mentioned properties are present both in species with a functional stinging apparatus (e.g., the formidably stinking mandi-bular gland secretions in the ponerine, *Paltothyreus tarsatus*, with its methyl sulfides, Casnati *et al.*, 1967) and in those with sting re-duction.

Jessen's gland in the ponerine, *Leptogenys ocellifera*, with its strongly smelling secretion, has to be considered an additional fighting organ, besides the mandibles and the poison gland.

The reductions and modifications in the stinging apparatus and its glands as described earlier are easily explained with respect to quick discharge and surface action. In the sting glands new venoms developed that were already efficient in the insect cuticle, and that no longer needed to be injected. Correspondingly, the injection mechanism was reduced and modified so that, at the end of the development, the secretions, without pumping, could be instantly applied to the body surface of the opponent. As was shown, both the poison gland (Formicinae, *Pheidole fallax*) and Dufour's gland (*Cremastogaster scutellaris*) are able to produce the new kind of secretions. In other instances another gland, such as Janet's gland, may take over the new combat function, and the stinging apparatus together with its glands is reduced (e.g., Dolichoderinae). In many species of Dolichoderinae the strongly insecticidal secretions point to specific venoms for specific enemies.

With these developments, however, the evolution of "new weapons" need not be terminated. They themselves may be superseded or replaced by other glands or fighting mechanisms. Examples, therefore, include the slavemaking *Polyergus* species, whose formic acid poison gland is considerably reduced. The mandibles in this genus have been specialized for the fights during slave raids (Huber, 1810; Forel, 1920; Stumper, 1960).

In other Formicinae the modified sting glands are superseded by strongly developed mandibular glands, as in *Dendrolasius fuliginosus* or the bursting *Colobopsis* species. In the dulotic *Formica pergandei* and *F. subintegra*, the Dufour's glands are strongly hypertrophied and repress the importance of the poison glands (Figure 4.7). The acetate secretions discharged during fights with defending workers of raided slave species colonies disperse and disorient the opponents, which thus can be easily overwhelmed (Regnier and Wilson, 1971). The function of superficially active poison secretions of the sting apparatus has been replaced by mechanically active sticky secretions in the myrmicine, *Cremastogaster inflata*, which primarily opposes the secretions of the hypertrophied metapleural glands (Figure 4.8).

Also, in *Pheidole biconstricta* (Figure 4.7) the strong development of Janet's gland as a fighting organ must have been preceded by another "new" fighting mechanism, since a reduction of the stinging apparatus took place not only in this but in all *Pheidole* species. The fighting mechanisms in ants thus exhibit a variety of development that seems unique in the animal kingdom.

New fighting techniques correspond with the modified or newly developed combat glands, for example, among the Dolichoderinae. *Tapinoma erraticum* during fights jump on the opponent and smear it quickly with the content of Janet's gland. *Technomyrmex albipes* proceeds toward the opponent and smears the toxic defensive secretion on it with lateral gaster movements. A similar fighting behavior has been observed in *Monomorium pharaonis* (Levinson *et al.*, 1974), which, however, still operates its pumping mechanism to apply the insecticide poison secretion superficially on the opponent. *Formica* often grasp the opponents with their mandibles for a short time before they smear or spray formic acid on to them. The lipophilic secretions of the Dufour's glands, which may act as carriers, enhance the surface action of the formic acid (Regnier and Wilson, 1969).

Perhaps the lack of a postpetiole in both groups, the *Formicinae* and *Dolichoderinae*, is connected with the development of new combat glands; a postpetiole may not be necessary since an exact pointing with the gaster in order to hit the opponent is not required. As was already mentioned above, the weak cuticle of many *Dolichoderinae* and *Formicinae* should be a corollary of the new fighting techniques.

Size of Colonies, Their Polymorphism, and Their Defensive Implications

It is obvious that species with generally small colonies, such as many Ponerinae, the myrmicines of the genus *Myrmecina*, and all *Leptothoracini*, which rarely have more than 300 adults in a nest, normally cannot develop a specialized defensive subcaste. An exception is the small colony of *Aneuretus simoni* (Aneuretinae) with rarely more than 100 adults, nevertheless possessing a dimorphic worker caste (Wilson *et al.*, 1956). However, a defensive function of the major workers was not observed.

The emergence of a soldier subcaste, in the form either of particularly large workers of the ordinary shape, or of morphologically distinct individuals, as in *Colobopsis*, parallels the evolution of a higher degree of worker polymorphism, with the colonies becoming larger (for details see Wheeler, 1910; Wilson, 1953, 1971).

Large colonies with numerous workers have, in addition to higher degrees of polymorphism, the potential for developing more sophisticated defensive systems. However, we stress here again that characteristics that actually enhance the defensive capacity of a society have not necessarily evolved due to the selective pressures of different menaces. Defensive power often seems to be a mere corollary of certain developments that are due to other selective pressures. In other words, defense is only one out of several factors responsible for the evolution of certain characters. Further consideration reveals that contradictory effects may often result from a given development.

As an impressive example we note the general evolutionary trend from small to large colonies. The more numerous the workers become, the higher the defensive power against aggressors or predators as well as against abiotic impacts. A large colony will be able to dig a deeper nest into the soil or wood, or, as in *Formica* species, build a larger thatchwork mound with better isolating capabilities. On the other hand, the larger a nest or a society, the more conspicuous it will become, the easier it will be uncovered by predators, and the more profitable it will be for a predator to specialize in searching for such colonies.

A second widespread evolutionary trait among ants is the increase in queen–worker dimorphism and worker polymorphism with higher specialization. In primitive species with small colonies the workers often are only slightly smaller than their queens, whereas in many highly evolved species groups considerable differences in the sizes of the two castes occur. Many authors claim that the evolution of female polymorphism in ants began with a primarily monomorphic worker caste with individuals rather large compared with their queens. The second step is a worker polymorphism due to allometric growth, which may develop toward a di- or triphasic allometry, or to a complete dimorphism of the worker caste (Wilson, 1953, 1971). Most of the species with a separate soldier subcaste, such as *Pheidole, Paracryptocerus*, and *Colobopsis*, may be classified among the latter. Finally, many of the most advanced genera have a secondarily monomorphic worker caste with considerable size differences between queens and workers. The larger workers apparently have been eliminated during the course of evolution.

There exist some species where tiny workers are clearly advantageous with respect to the specialized habits of gaining food. Thus, the thief ant, *Diplorhoptrum fugax*, has workers of only 1.4 to 2.5 mm length and queens which measure 5 to 7 mm. The small workers are able to reach the brood chambers of larger ants, where they prey on the immatures with the aid of their repellent poison gland secretion (see above), through narrow galleries where the owners of the brood cannot follow them. Similarly, the extreme queen–worker dimorphism can be explained in the genus *Carebara*, the members of which live as thief ants preying on termites (Wilson, 1971).

However, the production of small workers in great numbers, from the energetic point of view, seems to be advantageous in general, since small workers require less energy and can be produced in greater numbers. Numerous small workers are able to control the feeding area much better and more completely than a few larger ones, and they can recruit nest mates with their communication systems if help is needed, for example when a large food source has been found. In terms of defense, numerous small workers are likewise an advantage. One tiny worker, if captured by a predator, is a much smaller loss to the colony than one of only a few large workers. In addition, a small and often well-armed or heavily armored worker is an unattractive prey to many predators.

With this in mind, the formation of a worker caste on the whole must be considered as a means of defense for the more valuable reproductive queen. Apparently, in correlation with the evolution of smaller workers, the queens become absolutely larger and store more reserves in their wing musculature and fat bodies. This enables the queens of many species of the more advanced subfamilies to found new colonies in the so-called claustral manner (Forel, 1920; Wheeler, 1933; Wilson, 1971; and others). Queens of some genera (e.g., *Camponotus, Messor,* or *Lasius*) withdraw after the nuptial flight into a completely closed chamber in the soil or in wood, where they raise their first few workers by feeding the larvae exclusively with glandular secretions and eggs. Unlike queens of more primitive groups, like Myrmeciinae, most Ponerinae and some Myrmicinae, who sometimes leave the foundation chamber for foraging, the claustral founding queens remain well protected in their chambers. Only the first workers go out to collect food. Protection seems to be the main selective advantage of this kind of colony foundation.

Subterranean Life Habits

With respect to food habits, a predatory life is believed to represent the original state among ants (e.g., Wilson, 1971; Taylor, 1978). The evolution of trophobiotic relationships with plant lice opened a new source of food that was much less dangerous to collect than preying on other arthropods. Moreover, many ant species became specialized in tending plant lice in subterranean chambers, on roots, and thus could begin a more or less completely anachoretic existence, well protected in the soil. However, predator ant species with predominantly subterranean life habits are also known (e.g., *Ponera* and many other ponerines Leptanillinae, many Old and New World Dorylinae, Schneirla, 1971; Wilson, 1971; Gotwald, 1976). Predominantly nocturnal activity, which is reported in many ant species, especially in the tropics, may also be considered a means of defense, for example, against visually orientating predators. However, there seems to be no straight evolutionary direction toward nocturnal foraging, presumably because other factors, such as temperature in different climatic zones, must have a more direct influence on the behavior of the ants.

Polygyny, Polydomy, and Supercolonies

As in other social Hymenoptera, monogyny, the presence of only one queen in one nest, is considered to represent the "normal," and original condition. Modern genetic theories, especially the kinship selection theory, strongly support this opinion (Hamilton, 1964; Wilson, 1971, 1975c; Hölldobler and Wilson, 1977b). On the other hand, however, numerous ant species have been found to be regularly, or at least facultatively polygynous. According to Buschinger (1974) they represent about half the species from which reliable information can be obtained. Among the polygynous species we find many examples as the most successful ants ecologically.

Thus, *Formica* species very often have polygynous colonies which, by budding and branching, form supercolonies of numerous nests that remain in "amicable" relations, exchange food and nest mates and so on. Such species, for example, *Formica yessensis* in Japan with a supercolony of about 45,000 nests in Hokkaido (Higashi and Yamauchi, 1979), often represent the dominating ant species in their area. Also, some of the most important ant pests, for example, *Monomorium pharaonis* and *Iridomyrmex humilis*, are polygynous and polydomous; they even may be "unicolonial" in the sense of Hölldobler and Wilson (1977b) which means that there are no intraspecific boundaries between colonies or supercolonies of a given species. But what about the defensive aspects of monogyny and polygyny?

As already indicated, the formation of supercolonies lowers the intraspecific competition and enhances the potential of interspecific competiton, with the effect that competing species may be excluded from the supercolonies' territory (Higashi and Yamauchi, 1979). Many newly fertilized young queens are accepted within polygynous colonies of their own species, another means of reducing intraspecific competition, and colony foundation by budding lowers considerably the risks of independent foundation by single queens.

On the other hand, polygyny implies the danger that some degeneration of queen independence could occur (Brian, 1965). Often the queens of facultatively polygynous species are smaller than those of closely related monogynous ones, and thus are provided with less reserve substances and are less able to found their colonies independently. Under certain circumstances the evolution of social parasites from such polygynous parental species seems possible (Buschinger, 1970; Elmes, 1978).

In fact, most social parasitic species exhibit relationships rather similar to polygyny, being either polygynous themselves or parasitic toward polygynous host species (Buschinger, 1970). Monogyny, which is often secured by severe aggressiveness of queen and workers against additional queens in the nest, can therefore be understood as a defensive mechanism against the evolution of new, and the impact of existing, social parasites, which however, has been overcome again by a few social parasitic species whose queens kill the host colony queen and replace her (for example, *Lasius umbratus, Bothriomyrmex* species). As in so many examples described above, we encounter here again, with respect to the defensive significance of monogyny and polygyny, somewhat different strategies with contradictory results. Which strategy is adopted by a given species seems to depend on the relative importance of various selective pressures, and the defensive system of a species in most cases apparently reflects a compromise with regard to various requirements.

REFERENCES

Baroni-Urbani, C. (1979). Territoriality in social insects. In *Social Insects* (H. R. Hermann, ed.), Vol. I, pp. 81–120. Academic Press, New York.

Baroni-Urbani, C., and P. B. Kannowski. (1974). Patterns in red imported fire ant settlement of a Louisiana pasture: Some demographic parameters, interspecific competition and food sharing. *Environ. Entomol.* 3, 755–760.

Bazire-Bénazet, M., and L. Zylberberg. (1979). An integumentary gland secreting a territorial marking pheromone in *Atta* sp.: Detailed structure and histochemistry. *J. Insect. Physiol.* 25, 751–765.

Bergström, G., and J. Löfqvist. (1971). *Camponotus ligniperda* Latr. A model for the composite volatile secretions of Dufour's gland in formicine ants. In *Chemical Releasers in Insects* (A. S. Tahori, ed.), Proc. 2nd Int. IUPAC Congr. Pest. Chem., Tel Aviv, Vol. 3, pp. 195–223. Gordon and Breach, New York.

Bernardi, R., C. Cardani, D. Ghiringella, A. Selva, A. Baggini, M. Pavan. (1967). On the components of secretion of mandibular glands of the ant

Lasius (*Dendrolasius*) *fuliginosus. Tetrahedron Lett.* 40, 3893–3896.

Blancheteau, M. (1975). Observations sur le comportement agressif de la fourmi *Messor barbara* L. *Insecte Sociaux* 22, 113–116.

Blum, M. S., and H. R. Hermann. (1978a). Venoms and venom apparatuses of the Formicidae: Myrmeciinae, Ponerinae, Dorylinae, Pseudomyrmecinae, Myrmicinae, and Formicinae. In *Handbuch der experimentellen Pharmakologie*, Vol. 48, *Arthropod Venoms* (S. Bettini, ed.), pp. 801–869. Springer-Verlag, Berlin.

Blum, M. S., and H. R. Hermann. (1978b). Venoms and venom apparatuses of the Formicidae: Dolichoderinae and Aneuretinae. In *Handbuch der experimentellen Pharmakologie*, Vol. 48, *Arthropod Venoms* (S. Bettini, ed.), pp. 871–894. Springer-Verlag, Berlin, Heidelberg.

Blum, M. S., F. Padovani, A. Curley, and R. E. Hawk. (1969). Benzaldehyde: Defensive secretion of a harvester ant. *Comp. Biochem. Physiol.* 29, 461–465.

Bradshaw, J. W. S., R. Baker, and P. E. Howse. (1975). Multicomponent alarm pheromones of the weaver ant. *Nature* 258, 230–231.

Brand, J. M., M. S. Blum, H. A. Lloyd, and D. J. C. Fletcher, (1974). Monoterpene hydrocarbons in the poison gland secretion of the ant Myrmicaria natalensis (Hym., Form.). *Ann. Entomol. Soc. Am.* 67, 525–526.

Breen, J. A. G. (1976). Studies on Formica lugubris Zetterstedt in Ireland (Hymenoptera:Formicidae). Ph.D. Thesis, Nat. Univ. of Ireland, Dublin (cited in Baroni-Urbani, 1979).

Brian, M. V. (1965). *Social Insect Populations.* Academic Press, New York.

Brough, E. J. (1978). The multifunctional role of the mandibular gland secretion of an Austrialian desert ant, *Calomyrmex* (Hymenoptera: Formicidae). *Z. Tierpsychol.* 46, 279–297.

Brown, W. L., Jr. (1967). A new Pheidole with reversed phragmosis (Hymenoptera:Formicidae). *Psyche* 74, 331–339.

Brown, W. L., Jr. (1968). An hypothesis concerning the function of the metapleural glands in ants. *Am. Natur.* 102, 188–191.

Brown, W. L., Jr. (1973). *A Comparison of the Hylean and Congo–West African Rain Forest Ecosystems in Africa and South America: A Comparative Review*, (B. J. Meggers, E. S. Ayensu, and W. D. Duckworth, eds.). Smithsonian Inst. Press, Washington, D.C., 161–185.

Brown, W. L., Jr. (1976). Contributions toward a reclassification of the Formicidae. Part VI. Ponerinae, tribe Ponerini, subtribe Odontomachiti. Section A. Introduction, subtribal characters. Genus Odontomachus. *Studia Entomol.* 19, 67–171.

Brown, W. L., Jr., and W. W. Kempf. (1967). Tatuidris, a remarkable new genus of Formicidae (Hymenoptera). *Psyche* 74, 184–190.

Brown, W. L., and E. O. Wilson. (1959). The evolution of the dacetine ants. *Q. Rev. Biol.* 34, 278–294.

Brown, W. L., and B. P. Moore. (1979). Volatile secretory products of an Australian formicine ant of the genus *Calomyrmex* (Hymenoptera: Formicidae). *Insect Biochem.* 9, 451–460.

Buschinger, A. (1970). Neue Vorstellungen zur Evolution des Sozialparasitismus und der Dulosis bei Ameisen (Hym., Formicidae). *Biol. Zbl.* 88, 273–299.

Buschinger, A. (1972). Giftdrüsensekret als Sexualpheromon bei der Ameise Harpagoxenus sublaevis. *Die Naturwissensch.* 59, 313–314.

Buschinger, A. (1974). Monogynie und Polygynie in Insektensozietäten. In *Sozialpolymorphismus bei Insekten* (G. H. Schmidt, ed.), pp. 862–896. Wiss. Verlags GmbH, Stuttgart.

Buschinger, A. (1975). Sexual pheromones in ants. In *Pheromones and Defensive Secretions in Social Insects* (C. Noirot, P. E. Howse, and G. Le Masne, eds.). Proc. Symp. IUSSI Dijon, France, pp. 225–233.

Buschinger, A. (1976). Giftdrüsensekret als Sexualpheromon bei der Gastameise Formicoxenus nitidulus (Nyl.) (Hym., Form.). *Insecte Sociaux* 23, 215–226.

Buschinger, A., W. Ehrhardt, and U. Winter. (1980). The organization of slave raids in dulotic ants—A comparative study (Hymenoptera:Formicidae). *Z. Tierpsychol.* 53, 245–264.

Cammaerts, M.-C., E.-D. Morgan, and R. Tyler. (1977). Territorial marking in the ant Myrmica rubra L. (Formicidae). *Biol. Behav.* 2, 263–272.

Cammaerts-Tricot, M.-C. (1974). Piste et phéromone attractive chez la fourmi Myrmica rubra. *J. Comp. Physiol.* 88, 373–382.

Caroll, C. R., and D. H. Janzen. (1973). Ecology of foraging by ants. *Annu. Rev. Ecol. Syst.* 4, 231–257.

Casnati, G., A. Ricca, and M. Pavan. (1967). Sulla secrezione difensiva delle glandole mandibolari *Paltothyreus tarsatus* (Fabr.). *Chimica Ind.* 49, 57–61.

Cavill, G. W. K., and P. L. Robertson. (1965). Ant venoms, attractants, and repellents. *Science* 149, 1337–1345.

Chadha, M. S., T. Eisner, A. Monro, and J. Meinwald. (1962). Defense mechanisms of Arthropods. VII. Citronellal and citral in the mandibular gland secretion of the ant *Acanthomyops claviger* (Roger). *J. Insect Physiol.* 8, 175–179.

Collart, A. (1925). Quelques observations sur les fourmis Megaponera. *Bull. Cercle Zool. Cong.* 2, 26–28.

Coyle, F. A. (1966). Defensive behavior and associated morphological features in three species of the ant genus *Paracryptocerus. Insecte Sociaux* 13, 93–104.

Crawley, W. (1909). Queens of *Lasius umbratus* Nyl. accepted by colonies of *Lasius niger* L.. *Entomol. Month. Mag.* 20, 94–99.

Creighton, W. S. (1950). The ants of North America. *Bull. Mus. Comp. Zool.* 104, 1–585.

Creighton, W. S., and M. P. Creighton. (1959). The habits of *Pheidole militicida* Wheeler (Hymenoptera:Formicidae). *Psyche* 66, 1–12.

Crewe, R. M., and M. S. Blum. (1972). Alarm pheromones of the Attini: Their phylogenetic significance. *J. Insect Physiol.* 18, 31–42.

Cross, J. H., R. C. Byler, U. David, R. M. Silverstein, S. W. Robinson, P. M. Baker, J. S. Oliveira, A. R. Jutsum, and J. M. Cherrett. (1979). The major component of the trail pheromone of the leaf-cutting ant, *Atta sexdens rubropilosa*, Forel 3-ethyl-2,5-dimethylpyrazine. *J. Chem. Ecol.* 5, 187–203.

De Bruyn, G. J., and A. A. Mabelis. (1972). Predation and aggression as possible regulatory mechanisms in Formica. *Ecol. Pol.* 20, 93–101.

De Vroey, C. (1979). Aggression and Gause's law in ants. *Physiol. Entomol.* 4, 217–222.

Dobrzańska, J. (1978). Problem of behavioral plasticity in slave-making amazon-ant *Polyergus rufescens* Latr. and in its slave-ants *Formica fusca* L. and *Formica cinerea* Mayr. *Acta Neurobiol. Exp.* 38, 113–132.

Dobrzański, J. (1966). Contribution to the ethology of *Leptothorax acervorum* (Hymenoptera:Formicidae). *Acta Biol. Exp.* (*Warsaw*) 26, 71–78.

Duffield, R. M., J. W. Wheeler, and M. S. Blum. (1980). Methyl anthranilate: Identification and possible function in *Aphaenogaster fulva* and *Xenomyrmex floridanus*. *Flor. Entomol.* 63, 203–206.

Dumpert, K. (1978). *Das Sozialleben der Ameisen*. Verlag Paul Parey, Berlin/Hamburg.

Edmunds, M. (1974). *Defence in Animals*. Longman Group Ltd., Harlow, England.

Eibl-Eibesfeldt, I., and E. Eibl-Eibesfeldt. (1968). The workers bodyguard. *Animals* 11, 16–17.

Elmes, G. W. (1978). A morphometric comparison of three closely related species of Myrmica (Formicidae), including a new species from England. *Syst. Entomol.* 3, 131–145.

Evans, H. E., and M. J. W. Eberhard. (1970). *The Wasps*. David and Charles, Newton Abbot, England.

Foerster, E. (1912). Vergleichend-anatomische Untersuchungen über den Stachelapparat der Ameisen. *Zool. Jb. Anat. Ontog.* 34, 347–380.

Forel, A. (1874). *Les Fourmis de la Suisse*. Société Helvétique des Sciences Naturelles, Zürich.

Forel, A. (1878). Der Giftapparat und die Analdrüsen der Ameisen. *Z. Wiss. Zool.* (*Suppl.*) 30, 28–68.

Forel, A. (1920). *Les Fourmis de la Suisse*, 2nd ed. Impr. Coopérative, La Chaux-de-Fonds, Switzerland.

Forel, A. (1948). *Die Welt der Ameisen*. Rotapfel, Zürich.

Gabba, A., and M. Pavan. (1970). Researches on trail and alarm substances in ants. In *Advances in Chemoreception* (J. W. Johnston, Jr., D. G. Moulton, and A. Turk, eds.), Vol. 1. Appleton-Century-Crofts, New York.

Gamboa, G. J. (1975). Ant carrying in the desert leaf-cutter ant *Acromyrmex versicolor versicolor* (Pergande) (Hymenoptera:Formicidae). *Insectes Sociaux* 22, 75–82.

Ghent, R. L. (1961). Adaptive refinements in the chemical defense mechanisms of certain Formicinae. Ph. D. thesis, Cornell Univ., Ithaca, N.Y.

Gösswald, K. (1953). Histologische Untersuchungen an der arbeiterlosen Ameise Teleutomyrmex schneideri Kutter (Hym. Formicidae). *Mitt. Schweiz. Ent. Ges.* 26, 81–128.

Goetsch, W. (1934). Untersuchungen über die Zusammenarbeit im Ameisenstaat. *Z. Morphol. Ökol. Tiere.* 28, 219–401.

Goetsch, W. (1953). *Vergleichende Biologie der Insekten-Staaten*. Akad. Verl. Ges. Geest. u. Portig K. G., Leipzig, GDR.

Gotwald, W. H., Jr. (1976). Behavioral observations on African army ants of genus *Aenictus* (Hym., Form.). *Biotropica* 8, 59–65.

Gutmann, W. F., and K. Bonik. (1981). *Kritische Evolutionstheorie*. Gerstenberg Verlag, Hildesheim.

Hamilton, W. D. (1964). The genetical evolution of social behavior. I and II. *J. Theoretical Biol.* 7, 1–52.

Hermann, H. R. (1971). Sting autotomy, a defensive mechanism in certain social Hymenoptera. *Insecte Sociaux* 18, 111–120.

Hermann, H. R., and M. S. Blum. (1981). Defensive mechanisms in the social Hymenoptera. In *Social Insects* (H. R. Hermann, ed.), Vol. II, pp. 77–197. Academic Press, New York.

Hermann, H. R., and M. Douglas. (1976). Sensory structures on the venom apparatus of a primitive ant species. *Ann. Entomol. Soc. Amer.* 60, 681–686.

Higashi, S., and K. Yamauchi. (1979). Influence of a super-colonial ant *Formica* (Formica) *yessensis* Forel on the distribution of other ants in Ishikari Coast. *Jap. J. Ecol.* 29, 257–264.

Hölldobler, B. (1971). Sex pheromone in the ant *Xenomyrmex floridanus*. *J. Insect Physiol.* 17, 1497–1499.

Hölldobler, B. (1973). Chemische Strategie beim Nahrungserwerb der Diebsameise (*Solenopsis fugax* Latr.) und der Pharaoameise (*Monomorium pharaonis* L.). *Oecologia* 11, 371–380.

Hölldobler, B. (1976a). Recruitment behavior, home range orientation and territoriality in harvester ants, Pogonomyrmex. *Behav. Ecol. Sociobiol.* 1, 3–44.

Hölldobler, B. (1976b). Tournaments and slavery in a desert ant. *Science* 192, 912–914.

Hölldobler, B. (1977). Communication in social Hymenoptera. In *How Animals Communicate* (T. A. Sebeok, ed.), pp. 418–471. Indiana Univ. Press.

Hölldobler, B. (1978). Ethological aspects of chemical communication in ants. In *Advances in the Study of Behavior*, Vol. 8, pp. 75–115. Academic Press, New York.

Hölldobler, B. (1979). Territoriality in ants. *Proc. Am. Phil. Soc.* 123, 211–218.

Hölldobler, B., and H. Engel. (1978). Tergal and sternal glands in ants. *Psyche* 85, 285–330.

Hölldobler, B., R. Stanton, and H. Engel. (1976). A new exocrine gland in Novomessor (Hym., Form.) and its possible significance as a taxonomic character. *Psyche* 83, 32–41.

Hölldobler, B., and E. O. Wilson. (1977a). Colony-specific territorial pheromone in the African weaver ant *Oecophylla longinoda* (Latreille). *Proc. Natl. Acad. Sci. USA* 74, 2072–2075.

Hölldobler, B., and E. O. Wilson. (1977b). The number of queens: An important trait in ant evolution. *Naturwissenschaften* 64, 8–15.

Hölldobler, B., and E. O. Wilson. (1978). The multiple recruitment systems of the African weaver ant *Oecophylla longinoda* (Latreille) (Hymenoptera: Formicidae). *Behav. Ecol. Sociobiol.* 3, 19–60.

Howse, P. E. (1975). Chemical defences of ants, termites and other insects: Some outstanding questions. Symp. IUSSI Dijon, France, pp. 23–40.

Howse, P. E., R. Baker, and D. A. Evans. (1977). Multifunctional secretions in ants. Proc. VIIIth Int. Congr. IUSSI, Wageningen, 44–45.

Huber, P. (1810). *Recherches sur les Moeurs des Fourmis Indigènes.* J. J. Paschoud, Paris.

Jaffé, K., M. Bazire-Benazét, and P. E. Howse. (1979). An integumentary pheromone-secreting gland in Atta sp.: Territorial marking with a colony-specific pheromone in Atta cephalotes. *J. Insect Physiol.* 25, 833–839.

Janet, C. (1898). Système glandulaire tégumentaire de la Myrmica rubra: Observations diverses sur les fourmis. Carré et Naud, Paris.

Janzen, D. H. (1966). Coevolution of mutualism between ants and acacias in central America. *Evolution* 20, 249–275.

Janzen, D. H. (1967). Interaction of the bull's horn acacia (*Acacia cornigera* L.) with an ant inhabitant (*Pseudomyrmex ferruginea* F. Smith) in eastern Mexico. *Univ. Kansas Sci. Bull.* 47, 315–558.

Janzen, D. H. (1972). Protection of Barteria (Passifloraceae) by Pachysima ants (Pseudomyrmecinae) in a Nigerian rain forest. *Ecology* 53, 885–892.

Jentsch, J. (1969). Über das Giftsekret der Myrmicinen (For.). II. Anreicherung und Eigenschaften eines neuen Krampfgiftes aus der Ameise Myrmica ruginodis Nyl.. *Z. Angew. Zool.* 56, 337–356.

Jessen, K. (1977). Histologisch-morphologische Untersuchungen an abdominalen Hautdrüsen bei Leptogenys ocellifera und Odontomachus haematodes. Staatsexamensarbeit, Univ. Frankfurt.

Jessen, K., U. Maschwitz, and M. Hahn. (1979). Neue Abdominal-drüsen bei Ameisen. II. Ponerini. *Zoomorphologie* 94, 49–66.

Kneitz, G. (1970). Saisonale Veränderungen des Nestwärmehaushaltes bei Waldameisen in Abhängigkeit von der Konstitution und dem Verhalten der Arbeiterinnen als Beispiel vorteilhafter Anpassung eines Insektenstaates an das Jahreszeitenklima. *Verh. Deutsche Zool. Ges.* 64. (Tag.), 318–322.

Kugler, C. (1978). A comparative study of the myrmicine sting apparatus (Hymenoptera:Formicidae). *Studia Entomol.* 20, 413–548.

Kugler, C. (1979a). Evolution of the sting apparatus in the myrmicine ants. *Evolution* 33, 117–130.

Kugler, C. (1979b). Alarm and defense: A function for the pygidial gland of the myrmicine ant, Pheidole biconstricta. *Ann. Entomol. Soc. Am.* 72, 532–536.

Kunkel, H. (1973). Die Kotabgabe von Aphiden (Aphidina, Hemiptera) unter Einfluss von Ameisen. *Bonn. Zool. Beitr.* 24, 105–121.

Kutter, H. (1957). Eine neue Form der Sklavengewinnung bei Ameisen. *Umschau* 1957, 327–329.

Law, J. H., E. O. Wilson, and J. A. McCloskey. (1965). Biochemical polymorphism in ants. *Science* 149, 544–546.

Ledoux, A. (1950). Recherche sur la biologie de la fourmi fileuse (*Oecophylla longinoda* Latr.). *Ann. Sci. Nat. Zool.* (11ᵉ série), 12, 313–461.

Le Masne, G. (1965). Les transports mutuels autour des nids de *Neomyrma rubida* Latr.: Un nouveau type de relations (inter-spécifiques) chez les fourmis? CRV Congress UIEIS Toulouse, France, 303–322.

Leuthold, R. H., and U. Schlunegger. (1973). The alarm behaviour from the mandibular gland secretion in the ant *Crematogaster scutellaris*. *Insecte Sociaux* 20, 205–214.

Levieux, J., and T. Diomande. (1978). La nutrition des fourmis granivores. II. Cycle d'activité et régime alimentaire de *Brachyponera senaarensis* (Mayr) (Hymenoptera Formicidae). *Insecte Sociaux* 25, 187–196.

Levinson, H. Z., A. R. Levinson, and U. Maschwitz. (1974). Action and composition of the alarm pheromone of the bedbug *Cimex lectularius* L. *Naturwissenschaften* 61, 684–685.

Linsenmaier, W. (1972). *Insects of the World*. McGraw Hill, New York.

Löfqvist, J. (1976). Formic acid and saturated hydrocarbons as alarm pheromones for the ant *Formica rufa. J. Insect Physiol.* 22, 1331–1346.

Maidl, F. (1934). *Die Lebensgewohnheiten und Instinkte der staatenbildenden Insekten.* Verlag Fritz Wagner, Vienna.

Marikovsky, P. I. (1974). The biology of the ant *Rossomyrmex proformicarium* K. W. Arnoldi (1928). *Insecte Sociaux* 21, 301–308.

Markl, H. (1967). Die Verständigung durch Stridulationssignale bei Blattschneiderameisen. I. Die biologische Bedeutung der Stridulation. *Z. Vergl. Physiol.* 57, 299–330.

Markl, H., and S. Fuchs. (1972). Klopfsignale mit Alarmfunktion bei Rossameisen (Camponotus, Formicidae, Hymenoptera). *Z. Vergl. Physiol.* 76, 204–225.

Markl, H., and B. Hölldobler. (1978). Recruitment and food-retrieving behavior in *Novomessor* (Formicidae, Hymenoptera). II. Vibration signals. *Behav. Ecol. Sociobiol.* 4, 183–216.

Maschwitz, U. (1964). Gefahrenalarmstoffe und Gefahrenalarmierung bei sozialen Hymenopteren. *Z. Vergl. Physiol.* 47, 596–655.

Maschwitz, U. (1974). Vergleichende Untersuchungen zur Funktion der Ameisenmetathorakaldrüse. *Oecologia* 16, 303–310.

Maschwitz, U. (1975). Old and new chemical weapons in ants. Symposium IUSSI, Dijon, France. pp. 41–45.

Maschwitz, U., M. Hahn, and P. Schönegge. (1979). Paralysis of prey in Ponerinae ants. *Naturwissenschaften* 66, 213–214.

Maschwitz, U., K. Jessen, and E. Maschwitz. Foaming in Pachycondyla: A defence mechanism in ants. *Science* (in press).

Maschwitz, U., K. Koob, and H. Schildknecht. (1970). Ein Beitrag zur Funktion der Metathoracaldrüse der Ameisen. *J. Insect. Physiol.* 16, 387–404.

Maschwitz, U., and E. Maschwitz. (1974). Platzende Arbeiterinnen: Eine neue Art der Feindabwehr bei sozialen Hautflüglern. *Oecologia* 14, 289–294.

Maschwitz, U., and M. Mühlenberg. (1975). Zur Jagdstrategie einiger orientalischer Leptogenys-Arten (Form., Ponerinae). *Oecologia* 20, 65–83.

Maschwitz, U., and P. Schönegge. (1977). Recruitment gland of Leptogenys chinensis. *Naturwissenschaften* 64, 589.

Maschwitz, U. W. J., and W. Kloft. (1971). Morphology and function of the venom apparatus of insects—Bees, wasps, ants, and caterpillars. In *Venomous Animals and Their Venoms*, Vol. 3, pp. 1–60, Academic Press, New York.

McGurk, D. J., J. Frost, E. J. Eisenbraun, K. Vick, W. A. Drew, and J. Young. (1966). Volatile compounds in ants: Identification of 4-methyl-3-heptanone from *Pogonomyrmex* ants. *J. Insect Physiol.* 12, 1435–1441.

Möglich, M. H. J., and G. D. Alpert. (1979). Stone dropping by *Conomyrma bicolor* (Hymenoptera:Formicidae): A new technique of interference competition. *Behav. Ecol. Sociobiol.* 6, 105–113.

Olubajo, O., R. M. Duffield, and J. W. Wheeler. (1980). 4-heptanone in the mandibular gland secretion of the nearctic ant *Zacryptocerus varians* (Hymenoptera:Formicidae). *Ann. Entomol. Soc. Am.* 73, 93–94.

Parry, K., and E. D. Morgan. (1979). Pheromones in ants: A review. *Physiol. Entomol.* 4, 161–189.

Pavan, M. (1956). Studi sui Formicidae. II. Sull'origine, significato biologico e isolamento della dendrolasino. *Ricerca Scientifica* 26, 144–150.

Pavan, M., and G. Ronchetti. (1955). Studi sulla morfologia esterna e anatomia interna dell'operaia di *Iridomyrmex humilis* Mayr e richerche chimiche e biologiche sulla iridomyrmecina. *Atti Soc. Ital. Sci. Nat. Mus. Civ. Stor. Nat.* 94, 379–447.

Pickles, W. (1935). Populations, territory and interrelations of the ants *Formica fusca, Acanthomyops niger* and *Myrmica scabrinodis* at Garforth (Yorkshire). *J. Anim. Ecol.* 4, 22–31.

Pickles, W. (1936). Populations and territories of the ants, *Formica fusca, Acanthomyops flavus*, and *Myrmica ruginodis*, at Thornhill (Yorks.). *J. Anim. Ecol.* 5, 262–270.

Pontin, A. J. (1961). Observations on the keeping of aphid eggs by ants of the genus *Lasius* (Hym. Formicidae). *Entomol. Mon. Mag.* 96, 198–199.

Quilico, A., P. Grünanger, and M. Pavan. (1961). Sul componente odoroso del veleno del formicide Myrmicaria natalensis Fred. Proc. 11th Int. Congr. Entomol. Vienna, 1960. pp. 66–68.

Rathmayer, W. (1962). Das Paralysierungsproblem beim Bienenwolf Philanthus triangulum. *Z. Vergl. Physiol.* 45, 413–462.

Rathmayer, W. (1978). Venoms of Sphecidae, Pompilidae, Mutillidae and Bethylidae. In *Handbook of Experimental Pharmacology*, Vol. 48, *Arthropod Venoms*, (S. Bettini, ed.) pp. 661–690. Springer-Verlag, Berlin and New York.

Regnier, F. E., and E. O. Wilson. (1968). The alarm-defence system of the ant *Acanthomyops claviger*. *J. Insect Physiol.* 14, 955–970.

Regnier, F. E., and E. O. Wilson. (1969). The alarm-defense system of the ant *Lasius alienus*. *J. Insect Physiol.* 15, 893–898.

Regnier, F. E., and E. O. Wilson. (1971). Chemical communication and "propaganda" in slave-maker ants. *Science* 172, 267–269.

Reyne, A. (1954). *Hippeococcus*, a new genus of Pseudococcidae from Java with peculiar habits. *Zool. Mededelingen* 32, 233–257.

Rietschel, P. (1937). Bau und Funktion des Wehrstachels der staatenbildenden Bienen und Wespen. *Z. Morph. Ökol. Tiere* 33, 313–357.

Riley, R. G., R. M. Silverstein, and J. C. Moser. (1974a). Biological responses of *Atta texana* to its alarm pheromone and the enantiomer of the pheromone. *Science* 183, 760–762.

Riley, R. G., R. M. Silverstein, and J. C. Moser. (1974b). Isolation, identification, synthesis and biological activity of volatile compounds from heads of *Atta* ants. *J. Insect. Physiol.* 20, 1629–1637.

Ritter, F. J., I. E. M. Brüggeman-Rotgans, E. Verkuil, and C. J. Persoons. (1975). The trail pheromone of the Pharaoh's ant Monomorium pharaonis: Components of the odour trail and their origin. Proc. VIIIth Congr. IUSSI, Dijon, France, pp. 99–103.

Robinson, M. H. (1969). The defensive behaviour of some orthopteroid insects from Panama. *Trans. R. Entomol. Soc. London* 121, 281–303.

Schildknecht, H., and K. Koob. (1971). Myrmicacin, das erste Insekten-Herbizid. *Angew. Chem.* 83, 110.

Schmidt, G. H. (1961). Die physiologische Bedeutung des Puppenkokons von Ameisen. *Verh. XI. Int. Kongr. Entomol.* (*Vienna, 1960*) 1, 589–593.

Schmidt, J. O., and M. S. Blum. (1978). A harvester ant venom: Chemistry and pharmacology. *Science* 200, 1064–1066.

Schneider, P. (1971). Ameisenjagende Spinnen (Zodariidae) an Cataglyphis-Nestern in Afghanistan. *Zool. Anz.* 187, 199–201.

Schneirla, T. C. (1971). *Army ants: A Study in Social Organization*. (H. A. Topoff, ed.). W. H. Freemann, San Francisco.

Schumacher, A., and W. G. Whitford. (1974). The foraging ecology of two species of Chihuahuan desert ants: *Formica perpilosa* and *Trachymyrmex smithi neomexicanus* (Hymenoptera, Formicidae). *Insecte Sociaux* 21, 317–330.

Steiner, A. (1924). Über den sozialen Wärmehaushalt der Waldameise (Formica rufa var. rufo-pratensis For.). *Z. Vergl. Physiol.* 2, 23–56.

Steiner, A. (1929). Temperaturuntersuchungen in Ameisennestern mit Erdkuppeln, im Nest von Formica exsecta Nyl. und in Nestern unter Steinen. *Z. Vergl. Physiol.* 9, 1–66.

Stumper, R. (1960). Die Giftsekretion der Ameisen. *Naturwissenschaften* 47, 457–463.

Sudd, J. H. (1967). *An Introduction to the Behaviour of Ants*, E. Arnold Ltd., London.

Talbot, M. (1943). Population studies of the ant, *Prenolepis imparis* Say. *Ecology* 24, 31–44.

Taylor, R. W. (1978). *Nothomyrmecia macrops*: A living-fossil ant rediscovered. *Science* 201, 979–985.

Tumlinson, J. H., R. M. Silverstein, J. C. Moser, R. G. Brownlee, and J. M. Ruth. (1971). Identification of the trail pheromone of a leaf-cutting ant, *Atta texana. Nature* 234, 348–349.

Vowles, D. M. (1952). Individual behavior patterns in ants. *Adv. Sci* 10 (37), 18–21.

Wasmann, E. (1891). *Die zusammengesetzten Nester und gemischten Kolonien der Ameisen*. Münster, FRG.

Wasmann, E. (1934). Die Ameisen, die Termiten und ihre Gäste. Verlag G. J. Manz, A. G., Regensburg, Germany.

Wheeler, G. C., and J. Wheeler. (1979). Larvae of the social Hymenoptera. In *Social Insects* (H. R. Hermann, ed.), Vol. 1, pp. 287–338. Academic Press, New York.

Wheeler, J. W., S. L. Evans, M. S. Blum, and R. L. Torgerson. (1975). Cyclopentylketones: Identification and function in *Azteca* ants. *Science* 187, 254–255.

Wheeler, W. M. (1910). *Ants*. Columbia Univ. Press, New York.

Wheeler, W. M. (1922a). Ants of the American Museum Congo Expedition. A contribution to the myrmecology of Africa. VII. Keys to the genera and subgenera of ants. *Bull. Amer. Mus. Nat. Hist.* 45, 711–1004.

Wheeler, W. M. (1922b). Observations on *Gigantiops destructor* Fabricius and other leaping ants. *Biol. Bull. Wood's Hole* 42, 185–201.

Wheeler, W. M. (1933). *Colony Founding among Ants, with an Account of Some Primitive Australian Species*. Harvard Univ. Press, Cambridge, Mass.

Wheeler, W. M. (1936). Ecological relations of Ponerinae and other ants to termites. *Proc. Am. Acad. Arts Sci.* 71, 159–243.

Wilson, E. O. (1953). The origin and evolution of polymorphism in ants. *Q. Rev. Biol.* 28, 136–156.

Wilson, E. O. (1971). *The Insect Societies*. The Belknap Press of Harvard Univ. Press, Cambridge, Mass.

Wilson, E. O. (1975a). Enemy specification in the alarm-recruitment system of an ant. *Science* 190, 798–800.

Wilson, E. O. (1975b). *Leptothorax duloticus* and the beginnings of slavery in ants. *Evolution* 29, 108–119.

Wilson, E. O. (1975c). *Sociobiology*. The Belknap Press of Harvard Univ. Press, Cambridge, Mass., IX+697 p.

Wilson, E. O. (1976). The organization of colony defense in the ant *Pheidole dentata* Mayr (Hymenoptera:Formicidae). *Behav. Ecol. Sociobiol.* 1, 63–81.

Wilson, E. O., and W. H. Bossert. (1963). Chemical communication among animals. *Rec. Prog. Hormone Res.* 19, 673–713.

Wilson, E. O., T. Eisner, G. C. Wheeler, and J. Wheeler. (1956). *Aneuretus simoni* Emery, a major link in ant evolution. *Bull. Mus. Comp. Zool.* 115, 81–99.

Wilson, E. O., and F. E. Regnier, Jr. (1971). The evolution of the alarm-defense system in the formicine ants. *Am. Natur.* 105, 279–289.

Winter, U. (1979). Untersuchungen zum Raubzugverhalten der dulotischen Ameise Harpagoxenus sublaevis (Nyl.). *Insecte Sociaux* 26, 123–135.

Zakharov, A. A. (1969). The protection territory and development of red forest ant's colonies. *Probl. Forest Entomol.* 26, 187–199.

5
Morphology and Ultrastructure of Termite Defense Glands
André Quennedey

INTRODUCTION

A general review has been published on the defense mechanisms of termites (Deligne *et al.*, 1981). This chapter gives a more detailed study of the glands used directly for defense. However, such a study cannot be dissociated from the general topic of defense in termites; therefore, I invite the reader to refer to this review.

Within the last decade there has been a notable increase in knowledge of the chemical nature of termite defense secretions (cf. reviews of Prestwich, 1979; Deligne *et al.*, 1981). On the contrary, data on the ethological and morphological aspects of termite defense have remained fragmentary (cf. review of Noirot, 1969). In addition, the anatomical studies of the frontal defense glands of termites are often a half century or more old (Holmgren, 1909; Bugnion and Popoff, 1910; Bonneville, 1936). Recent articles in which scanning and transmission electron microscopy were used are still few in number and unfortunately consider only a few termite species (Quennedey, 1973, 1975a, b, 1978; Quennedey and Deligne, 1975; Deligne and Quennedey, 1977; Deligne *et al.*, 1981). Because of this lack of information and the widespread nature of the subject, new unpublished observations on termite defense glands are included here. The relationship of these data to existing information on the chemistry of termite defense behavior is presented.

The soldier caste seems the most specialized caste to be found in all social insects. The soldiers appear to be highly specialized, not performing any work in the society and being unable to feed themselves (Grassé, 1939). From a morphological point of view, the soldier's head is highly sclerotized. More and more sophisticated defense systems have been found in this caste. In addition to phragmotic and mandibular weapons (see Deligne *et al.*, 1981) various glandular organs have been differentiated.

This chapter considers the soldier's glands in which the defensive role is well established and also those whose structure and position are in agreement with a probable defensive function. In some species the workers also assume a direct defensive function (because of a lack or small number of soldiers) by defecating mechanisms (*Skatitermes*, Coaton, 1971; *Speculitermes*, Sands, 1972) or by peculiar suicidal behavior (exploding gut of *Anoplotermes* and most Apicotermitinae, Sands, 1972). The role of the tarsal glands in the defense of workers (Bacchus, 1979) remains hypothetical. Some swarming alates, being distasteful to birds, present an individual defense; however, the possible glandular origin of such distastefulness has not been thoroughly studied.

EPIDERMAL GLANDS DIRECTLY ASSOCIATED WITH THE MANDIBLES OF SOLDIER TERMITES

The Cibarial Glands

In the soldiers of Macrotermitinae the labrum is well developed and overlaps at least half of the mandibles (Figures 5.11d and 5.12b). In many genera (some *Macrotermes, Acanthotermes, Pseudacanthotermes, Synacanthotermes*) the labrum has a swollen translucid apical part—the hyaline tip (Figure 5.1e and f). Such specialization was also observed in Apicotermitinae (*Apicotermes desneuxi*, Emerson, 1952a; *Foraminitermes*) and among the primitive genera of Nasutermitinae: *Syntermes* (Emerson, 1945) (Figure 5.1b), *Cornitermes* (Emerson, 1952b) (Figure 14c) and *Labiotermes* (Emerson and Banks, 1965).

In cross section, the epidermis facing the inner surface of the labrum is thickened with a glandular aspect and is surrounded by a sclerotized cuticle. A cross section of the hyaline tip shows a thicker cuticle composed mainly of unsclerotized and loose endocuticle, giving the hyaline appearance of the labral tip (Figure 5.1a, c, and d).

Such a glandular structure was originally called a labral gland (Deligne *et al.*, 1981), but the recent discovery of a similar secretory epithelium at the upper surface of the hypopharynx led us to consider both these glands as the same organ: the cibarial gland (Figures 5.1a and 5.12a). The hypopharyngeal part of the cibarial gland is cytologically very different from the hypopharyngeal glands observed in the different castes of the lower termites and which have disappeared in Rhinotermitidae and Termitidae (Brossut, 1973).

The position of the cibarial gland with respect to the mandibles suggests that the secretion may impregnate the mandibular blades and thus penetrate the wound inflicted by the mandibles. Although the chemical nature of the secretion remains unknown, we may consider this gland as an auxiliary defense gland whose possible action is to add to the toxic effect of the salivary secretion of the soldiers.

Only the glandular epithelium of the labral part of the cibarial gland of *Macrotermes bellicosus* was studied by electron microscopy (Figure 5.2a). It is composed of class 1 cells (according to the classification of Noirot and Quennedey, 1974). Each lengthened cell (40 μm long and 5 μm wide) has a dense cytoplasm containing an oval-shaped nucleus (8 $\times$ 4 μm) located in the basal region (hemocoel side), clear golgi bodies, mitochondria with an electron-dense matrix, aggregated glycogen granules, numerous microtubules arranged parallel to the length of the cell, and a smooth endoplasmic reticulum scattered in a vesicular form in the cytoplasm and in a tubular form along the lateral plasma membrane. A septate junction with an apical desmosome is seen in the upper part of the intercellular junction area. On the inner side of the gland facing the hemocoel, the plasma membrane is lined with a thin, highly folded basal lamina, suggesting that active fluid transport occurs. The opposite side facing the glandular cuticle shows an irregular microvillous border composed of tightly packed and narrow microvilli 2 μm in length with inner longitudinal microfilaments.

The cuticle lining the inner surface of the labrum is 8 μm thick and is composed of approximately 15 endocuticular laminae crossed by numerous pore canals and an outer thin epicuticle pitted with epicuticular pores. Between the cuticle and the surface of the cells is an extracellular space about 2 μm in depth filled with an abundance of dense material secreted by the cells. In the hyaline tip (Figure 5.2b), the cuticle is thicker (20 μm) but most of the endocuticular laminae have a loose structure, thus greatly enlarging the extracellular space (30 μm thick) that becomes filled with the glandular secretion. This secretion crosses the cuticle by way of epicuticular pores. The extracellular space in the hyaline tip appears to function as a reservoir.

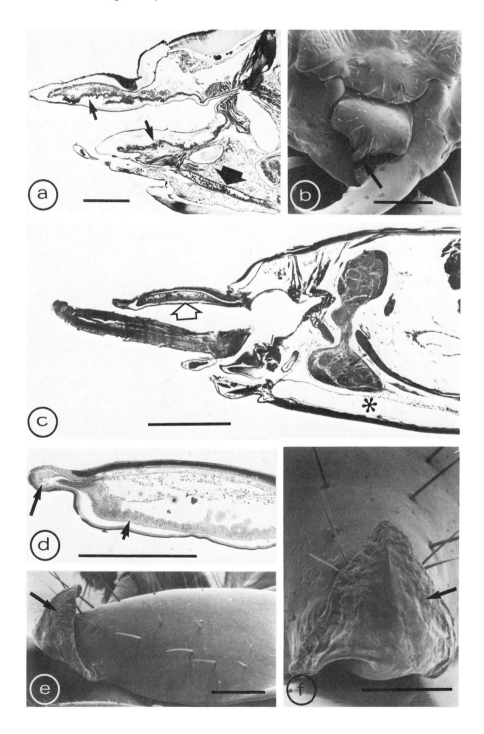

Epidermal Glands at the Base of Mandibles

Among the Apicotermitinae, soldiers of *Machadotermes* (Figure 5.3a) exhibit short and powerful mandibles with an enlarged, rounded but less sclerotized base (Weidner, 1974). Approximately 1000 sensory bristles (50–70 μm in length) and numerous small holes (2 μm in diameter) are observed on the surface of the base (Figure 5.3b). These holes correspond to the exits of the class 3 epidermal units according to the classification of Noirot and Quennedey (1974). Bristles and glandular openings are also present on the entire anterior part of the soldier's head. Around the fontanelle, the epidermal cell openings are often gathered in clusters of 6 to 10. On the mandibles, the openings are always isolated, but on the surface of the bulbous base, near the sharp tip of the mandibles, there are parallel cuticular furrows (Figure 5.3b, c). These furrows may facilitate and direct the discharge of glandular secretions toward the mandibular blades.

THE SALIVARY WEAPON

Whereas salivary glands of worker termites are involved with their feeding and building activities (Noirot, 1969), those of some soldiers have a defensive function. Such polyethism of the glands occurs without any major modification of its organization, the general structure of which is the same as that encountered in Orthoptera and Dictyoptera.

FIGURE 5.1. Cibarial glands. (a) Sagittal section of *Syntermes wheeleri* soldier head showing the glandular thickenings of the inner surface of the labrum and the upper surface of the hypopharynx (small arrows). The large arrow indicates the terminal part of a salivary reservoir. ×25—scale: 0.5 mm. (b) Scanning electron micrograph showing a frontal view of *Syntermes* soldier head. Note the hyaline tip (arrow) at the end of the labrum. ×16—scale: 1 mm. (c) A lateral sagittal section of *Macrotermes bellicosus* minor soldier. The labral part (arrow) of the cibarial gland and the anterior part of a salivary reservoir (asterisk) are seen. ×45—scale: 0.5 mm. (d) Magnified sagittal section of *M. bellicosus* major soldier's labrum. The thickening of the inner epithelium is clearly seen (small arrow) and the hyaline tip is mainly composed of a thick unsclerotized cuticle (large arrow). ×65—scale: 0.5 mm. (e and f) Lateral and frontal scanning electron micrographs of *M. bellicosus* major soldier's hyaline tip. The artifactual bending up of the tip shows the smooth appearance of the inner cuticle (arrows). (e) ×70; (f) ×120—scales: 0.2 mm.

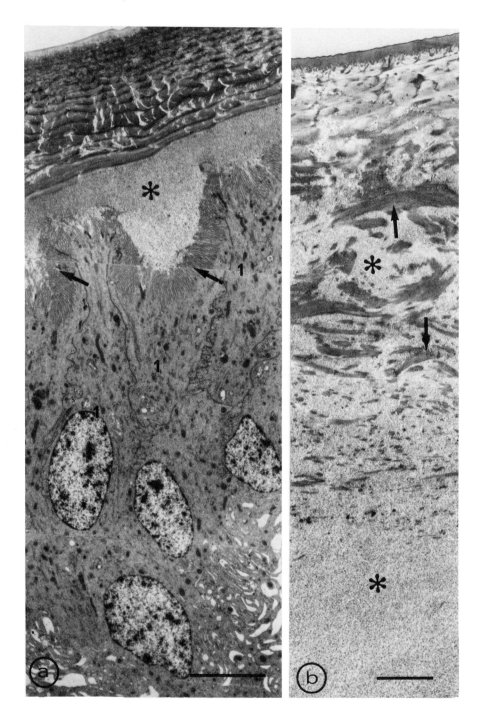

The labial or salivary glands are two symmetrical organs made of several clusters of lobular acini each composed of three types of glandular cells: vacuole cells and foam cells that are large in size, while the third type, the parietal cells, are small and situated at the periphery of the acini. From the acini, small salivary canals join together to form a collector canal that connects them with the base of the labium. Both these glands open into the preoral cavity. In addition, a salivary reservoir exists at the base of each collector canal.

Such a structure is found throughout all termite castes, but in the soldiers in which salivary glands have a defensive function, the acini are less developed (lateral and dorsal acini are reduced) and the salivary reservoirs are hypertrophied. They extend deep into the abdomen, pushing up and behind the digestive tract. In the primitive genus *Mastotermes* the reservoirs are more reduced and penetrate into the body to the depth of the second abdominal tergite. Among Termitinae, the salivary reservoirs of *Globitermes sulphureus* occupy the entire front half of the soldier's abdomen (Noirot, 1969). Among Macrotermitinae, the reservoirs extend from the base of the labium forward as far as half (*Microtermes alboparticus, Odontotermes smeathmani*) or two-thirds (*Allodontotermes giffardii, Hypotermes xenotermitis*) of the length of the abdomen (Figure 5.4a). In the large soldiers of *Pseudacanthotermes spiniger* the reservoirs occupy nine-tenths of the abdomen (Noirot, 1969).

The following description is of the ultrastructure of the salivary glands of *Macrotermes bellicosus* minor soldiers; however, both soldier types of this species have a similar glandular structure. Within the acini, the large glandular cells nearly surrounded the branched salivary canal to which they often connect in pairs. These cells have a large round nucleus (5 μm in diameter) that contains numerous aggregates of chromatin. From the appearance of the cytoplasm, two types of cells can be distinguished. These types may correspond to vacuolar and foam cells or they may represent different physiological

FIGURE 5.2. The cibarial gland of *Macrotermes bellicosus* minor soldier. (a) An electron micrograph showing the class 1 cells (1) with a dense cytoplasm and numerous apical microvilli (arrows). Between them and the thick sclerotized cuticle an extracellular space filled with glandular secretion (asterisk) is seen. ×4,000—scale: 5 μm. (b) In the hyaline tip, the cuticle is four times thicker but loosely arranged (arrows) and a great amount of secretion (asterisks) is present in this extracellular reservoir. ×6,000—scale: 2.5 μm.

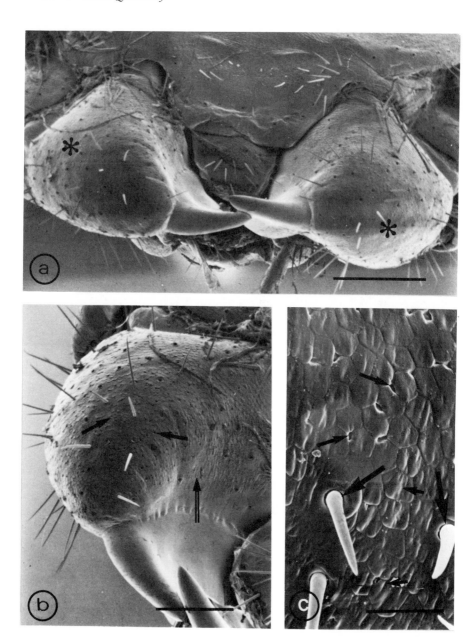

FIGURE 5.3. Scanning micrographs of the epidermal glands of *Macha-dotermes* soldier mandibles. (a) Frontal view showing the peculiar enlarged base of mandibles (asterisks). ×125—scale: 0.2 mm. (b) On the bulbous part of the mandible, sensory bristles and glandular openings (single arrows) are ob-served. Near the mandibular blade, cuticular furrows are differentiated (twin

states of the same cell. More often, the rough endoplasmic reticulum is extensive, giving a striated appearance to the cytoplasm; mitochondria are lengthened and the golgi apparatus, which secretes large clear vesicles, is well developed. Some cells that are less abundant and reduced in size have a clearer cytoplasm containing free ribosomes, more rounded mitochondria, and a less developed golgi apparatus (Fig. 4b).

Toward the center of the acini, the apical plasma membrane forms radially arranged, tightly packed long microvilli that take up all the space of the extracellular cavity related to the salivary canal. The intercellular space has a twisted septate junction. At the periphery, the membrane is greatly folded, increasing the surface contact of the cells with the hemocoel. Here and there, small and flattened parietal cells with oval nuclei and clear cytoplasm are observed. A thick (2 μm) and dense basal lamina, which is penetrated by tracheola and neurosecretory axons, entirely overlaps the acini.

The salivary canal has a nearly circular cross section as it emerges from the acini, and the cells surrounding the canal lumen have a regular shape, giving the canal its tubular structure (Figure 5.5a). At the beginning, within the acini, the cells are irregular and extensively overlapped, as indicated by the convoluted intercellular junctions (Figure 5.4b). In these cells the nucleus is generally lengthened and irregular with a dense nucleoplasm, while the cells of the free canal have more rounded nuclei. In both cases the cytoplasm is particularly rich in glycogen and mitochondria. The plasma membrane facing the canal lumen is formed into a tightly packed layer of microvilli covered by an irregular layer of epicuticle. The cells of the collector canal are characterized by well-differentiated infoldings of their basal membrane between which mitochondria are seen (Figure 5.5b and c). These infoldings have equidistant opposed plasma membranes; such a structural feature is reminiscent of the so-called scalariform junctions described in the insect epithelia as carrying ions and water (Fain-Maurel and Cassier, 1972; Noirot-Timothée *et al.*, 1979). In addition, the epicuticle surrounding the lumen is thicker and composed of inner and outer layers. The epicuticular surface appears regularly undulating in cross section because of the taenidial organization of the duct limiting the lumen of the collector canal. Thus the salivary canals cannot simply be considered as conducting canals but

arrow). ×210—scale: 0.1 mm. (c) Magnified view of the bristles (large arrows), the glandular openings (medium arrows) and the furrows (small arrows). ×1,050—scale: 20 μm.

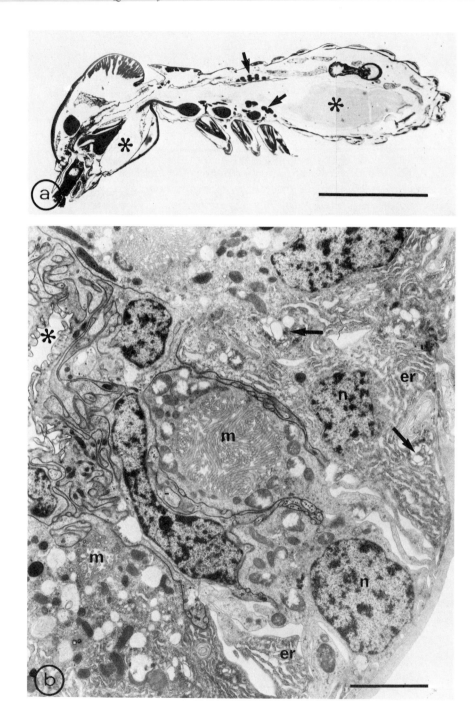

are also involved in the modification of the secretion (Noirot, 1969).

On the other hand the huge salivary reservoirs in which the secretion is stored are made up of a thin wall (about 1 μm thick) composed of very flattened cells (0.3 μm wide) that are a little thicker (1 μm) at the level of the nucleus (Figure 5.5d). The cytoplasm contains mainly microtubules, and a sinuous and lengthened septate junction is observed laterally. The apical and basal plasma membrane is regular and undifferentiated; a thin basal lamina faces the basal plasma membrane. The cuticle limiting the salivary reservoirs is composed of epicuticle (inner and outer) and a single endocuticular lamina. Neurosecretory innervation, with areas of synaptoid vesicles, is observed in the acini, the salivary canals, and the reservoirs at the level of the basal lamina.

Moore (1968) has isolated from the salivary glands of *Mastotermes* a mixture of 1,4-benzoquinone, toluquinone, and proteinaceous constituents. Either one or the other of the two quinones is also present in the labial secretions of termititid species (Maschwitz *et al.*, 1972; Maschwitz and Tho, 1974; Wood *et al.*, 1975; Howse, 1975). Nevertheless, *Macrotermes carbonarius* also produces in addition to 1,4-benzoquinone its methyl homolog (Maschwitz and Tho, 1974) and the main component of the *Ancistrotermes cavithorax* secretion is a furanosesquiterpenoid (Howse, 1975; Evans *et al.*, 1977).

THE FRONTAL WEAPON

Quennedey and Deligne (1975) have defined the frontal weapon as being composed of the frontal gland and the cephalic regions (frons, clypeus, and labrum) whose modified structure facilitates the discharge of the frontal secretion.

FIGURE 5.4.   The salivary glands. (a) Longitudinal section of *Hypotermes xenotermitis* soldier showing the great development of the salivary reservoirs (asterisks) pushing up the digestive tube. Several glandular acini (arrows) are observed in the metathorax. ×30—scale: 1mm. (b) Electron micrograph of a glandular acini of *Macrotermes bellicosus* minor soldier in cross section. The glandular cells are rich in rough endoplasmic reticulum (er), giving a striated appearance to the cytoplasm, and in golgi bodies (arrows). The nuclei (n) are situated at the periphery and the microvilli (m) in the center of the acini near the salivary canal (asterisk) are composed of contorted cells. ×7,000—scale: 3 μm.

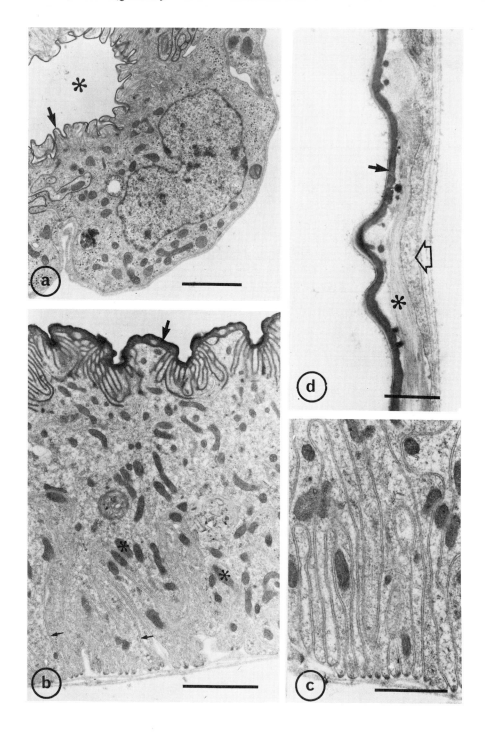

The frontal gland appears to be an unpaired organ differentiated from the cephalic epidermis of the median frontal region without equivalent in other insects (Noirot, 1969). However, although such an organ may be absent in many species of termites, where it does exist there are notable structural differences between the gland of the various castes, and also the gland within each caste. Generally in workers and imagoes the gland is a thickening of the frontal epidermis, as in *Cubitermes* (Deligne, 1970), and in this region the frontal cuticle shows a small but distinct hollow without a glandular opening. In *Macrotermes* imagoes, electron microscopy reveals a gland composed of class 1 cells with a podocyte-like base (unpublished observations). Cytological data about the glands of both castes (workers and imagoes) remain very poor, and the role of the gland is still open to question. In spite of the great differences between their structure and those of soldiers, a defensive role cannot be totally eliminated.

Thanks to Holmgren (1909), we have a better idea of the distribution of the gland in soldiers. Using several criteria, in particular the relative development of the frontal gland, he distinguished three systematic subdivisions among the termites. He regrouped, within the Protermitidae, the lower families (Mastotermitidae, Kalotermitidae, Termopsidae, and Hodotermitidae) in which the frontal gland is undifferentiated. In his other families, the Mesotermitidae (Rhinotermitidae) and the Metatermitidae (Termitidae), the frontal gland is generally present and well developed.

In all the soldiers studied the gland appears as an invagination of the frontal epithelium (the frontal pore), which extends into the soldier's body by various degrees. The Rhinotermitidae exhibit the most developed gland, which often extends as far as the abdomen, while in the Termitidae this organ is smaller and is always contained within the head capsule.

FIGURE 5.5.   The salivary glands of *Macrotermes bellicosus* minor soldier. (a) A cross section showing a quarter of a salivary canal at its emergence from the acini. Note the thin epicuticle (arrow) lining the salivary lumen (asterisk). ×8,000—scale: 2 μm. (b) Typical salivary canal with numerous basal infoldings (small arrows), abundant mitochondria (asterisks) and thicker epicuticle (large arrow) with a taenidial organization. ×10,000—scale 2 μm. (c) Higher magnification of the regularly arranged basal infoldings, such a feature being reminiscent of scalariform junctions. ×20,000—scale: 1 μm. (d) The salivary reservoir wall is particularly thin with very flattened cells (large arrow). The cuticle lining the reservoir is composed of epicuticle (small arrow) and a single endocuticular lamina (asterisk). ×30,000—scale 0.5 μm.

Rhinotermitidae

The family Rhinotermitidae consists of seven well-defined sub-families with many similar evolutionary characteristics, particularly at the glandular level (Ampion and Quennedey, 1980). The frontal gland is always present in soldiers and generally well developed; however, the modified cephalic structures related to the discharge of the frontal secretion have evolved in different directions (Quennedey and Deligne, 1975).

Coptotermitinae

The soldiers of the single genus *Coptotermes* constituting this subfamily secrete a white latex from a large triangular frontal pore that quickly dries, thereby entangling the soldier and its opponent. The large size of the frontal opening (300 μm) is necessary because of the viscosity of the secretion, which covers the clypeus and the mandibles. Chemical analysis has shown that it is a suspension of saturated *n*-alkanes (C_{22}–C_{27}) in an aqueous solution of mucopoly-saccharides composed of glucosamine and glucose units (Moore, 1968, 1969).

The frontal gland, which is very large, pushes back the digestive track and takes up almost the entire body of the soldier (Bugnion and Popoff, 1910). Such an extraordinary development seems to be a general rule in this genus; for example; in *C. intermedius* the frontal gland is 3 mm in length and 1 mm in diameter, which is almost the same size as the soldier. This insect could almost be thought of as a walking gland. In this case the glandular wall is a very thin (4 to 10 μm high) secretory epithelium. Under light microscopy the cells of the gland have a vacuolar appearance when fixed with alcohol.

With electron microscopy the gland has a pleated structure and is composed of a mixture of class 1 and class 3 cells (Quennedey, 1975c, 1978; Figure 5.6a). The class 1 cells are flattened and their cytoplasm contains a long and multilobed nucleus with a dense nucleoplasm, many mitochondria, small golgi bodies, free ribosomes, an abundance of aggregated glycogen granules, and numerous myeloid bodies, all of which may be considered as constituting the glandular secretion. The plasma membrane under the cuticle forms short (0.3–0.5 μm) microvilli with inner longitudinal microfilaments. The regular mem-brane facing the hemocoel is lined by a thin basal lamina. The class 3 units are composed of two cells: a secretory cell and a narrow canal cell. The secretory cells show a characteristic central extracellular space delimited by long microvilli (1 μm). The epicuticular end-apparatus found within the extracellular space is connected at one end to the cuticular duct of the canal cell (Figure 5.6b). In the cytoplasm of the secretory cell, there is extensive rough endoplasmic reticulum and large golgi bodies producing numerous clear vesicles with flocculent contents. Any intercellular junctions are seen in the narrow intercellular space between the class 1 and the secretory class 3 units. In contrast, between the secretory cells and the canal cells, and between the canal cells and the class 1 cells, a septate junction with an apical zona adhaerens is present. The flocculent secretion observed in the vesicles is also visible in the extracellular space of the secretory cells, in the lumen of the end-apparatus, and in the cuticular duct, before it is found stored in the broad reservoir of the frontal gland. The cuticle lining the reservoir is very thin (0.5–2 μm thick) and only composed of epicuticular layers (Figure 5.6b). Large electron-dense droplets secreted by the class 1 cells are seen between the outer epicuticle, bordered by the epicuticular filaments, and the inner epicuticle, bordered by the class 1 microvilli. This secretion is observed mixed with the flocculent material in the glandular reservoir.

The dual origin of the secretion is noteworthy. The flocculent phase is secreted by the class 3 units and the dense droplets by the class 1 cells. According to the histochemical test of Thiéry (1967) the flocculent material seems to be composed of saccharides, that, like glycogen, are able to bind silver granules (Figure 5.6c and d). It is interesting to compare this result with Moore's (1969) biochemical data. If it is possible that the flocculent secretion is in fact a polysaccharide, the dense droplets cannot be assimilated into *n*-alkane, such compounds being eliminated by the ultrastructural fixation procedures.

Psammotermitinae

The subfamily Psammotermitinae also contains only a single genus, *Psammotermes*. The soldiers are quite variable in size, but are usually divided into two classes, minor and major. In both of these, the frontal gland is well developed and has the same organization. It penetrates deep into the body to the depth of the third abdominal tergite, thus making the length of the gland about two-thirds the size of the soldier.

The gland opens into the middle of the frons by a small frontal pore (20 μm and 40 μm in diameter in minor and major soldiers, respectively). A median gutter, more pronounced in the minor soldier, facilitates the discharge of the secretion toward the clypeolabrum and the strong and denticulated mandibles (Quennedey and Deligne, 1975).

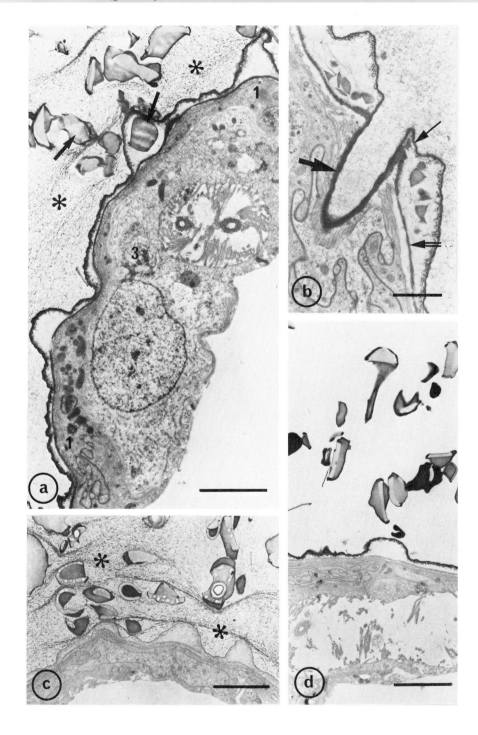

The thin (5 μm) glandular wall is composed of only class 1 cells (Figure 5.7a). Each cell overlaps its neighbors with an extremely sinuous lateral plasma membrane supported along its upper half by a septate junction and an apical zona adhaerens. The plasma membrane lying under the cuticle has an irregular surface but does not differentiate microvilli and is more regular on the side facing the basal lamina. The cytoplasm contains a nucleus, located in the basal region of the cell, mitochondria, smooth endoplasmic vesicles, glycogen, and numerous microtubules. Secretory globules with a dark electron-dense periphery and a clear core are abundant and are distributed throughout the region between the basal lamina and the cell and in the intercellular space (Figure 5.7b). These globules are also observed closely associated with the mitochondria within the cell. Finally they are recognizable on the upper surface of the plasma membrane under the cuticle. Such observations suggest that secretory precursors are collected from the hemocoel, channeled via the intercellular spaces, and brought into contact with the mitochondria before crossing to the cuticle and collecting in an amorphous state in the reservoir of the gland. The glandular cuticle is only composed of an outer epicuticle (15 nm thick) pitted with holes 30 nm in diameter. Epicuticular filaments are numerous and irregularly packed below it (Figure 5.7c).

Heterotermitinae

The related genera *Heterotermes* and *Reticulitermes* show many structural similarities of the head capsule of the soldiers. The small frontal pore (20 μm in diameter) is situated at the end of a longitudinal hollow in the frons. In addition, this hollow in *Reticulitermes* has several grooves within it, the orientation and the slope of which would appear to aid the flow of the secretion toward the sharp mandibles (Quennedey and Deligne, 1975).

FIGURE 5.6. The frontal gland of *Coptotermes lacteus* soldier. (a) The glandular epithelium is composed of class 1 cells (1) and class 3 units (3). In the gland lumen the secretion is composed of (i) large globules crossing the cuticle (arrows) and produced by class 1 cells and (ii) a flocculent material (asterisks) secreted by class 3 units. ×5,600—scale: 3 μm. (b) The glandular cuticle is only formed by separated inner (twin arrow) and outer epicuticle (small arrow) which reunite to form the canal duct (large arrow) of a class 3 unit. ×12,500—scale: 1 μm. (c and d) Results of the histochemical test of Thiéry (1967); (c) the flocculent material (asterisks) is silver labeled after periodic oxidation; (d) without oxidation this material is not visible; thus, its glucidic nature is demonstrated. ×5,000—scales: 3 μm.

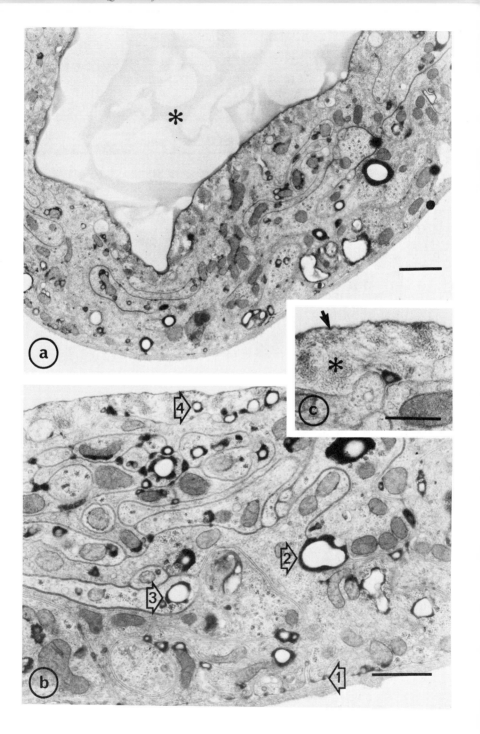

Reticulitermes lucifugus (Noirot, 1969) and *R. santonensis* are similar in that their frontal gland extends forward into the prothorax of the soldier, although its size remains relatively small (2.5 mm long and 0.2 mm in section) compared with the size of this gland in the genera discussed previously. The glandular epithelium is composed only of class 1 cells (10–20 μm thick), but at the level of the frontal pore it appears thicker (100 μm, Figure 5.8a) because of the presence of several class 3 units opening around the pore.

The cytoplasm of the class 1 cells contains a large nucleus (8 μm long) located basally (Figure 5.8b). The microtubules and the smooth endoplasmic reticulum are so abundant that they give the cytoplasm an electron-dense appearance. Spots of electron-dense secretion are observed in the upper part of the cell. The apical plasma membrane is differentiated into microvilli with inner longitudinal microfilaments. The intercellular junctions are very sinuous and have upper septate junctions and apical zona adhaerens. Facing the thin basal lamina, the plasma membrane has many long and narrow infoldings and numerous hemi-desmosomes where the microtubules are anchored.

The cuticle (Figure 5.8c) is composed of a deeper dense cuticular layer (0.3 μm in thickness) without a recognizable structure and an epicuticle with both inner and outer layers pitted by epicuticular pores. Two extracellular spaces full of a fine secretory material are observed beneath the deeper cuticular layer and also between it and the epicuticle. Below the epicuticle, the epicuticular filament zone has a depth of 0.5 μm.

Termitogetoninae

Only histological and scanning electron microscope data are available on the tropical genus *Termitogeton*. The soldiers have a characteristic flat, heart-shaped and hairy head. In the center of the frons a triangular area is clearly seen into which opens a narrow (3–5 μm) frontal pore; 60 to 80 small glandular openings (0.2 μm in diameter) are scattered in this area (Quennedey and Deligne, 1975).

FIGURE 5.7. The frontal gland of *Psammotermes hybostoma* soldiers. (a) The glandular epithelium is only composed of class 1 cells and in the lumen of the gland an amorphous secretion is present (asterisk). ×11,500—scale: 1 μm. (b) Abundant secretory droplets are observed in the tissue along the basal lamina (1), within the intercellular space (2), in the cell (3), and outside of the cell (4). ×16,500—scale: 1 μm. (c) Higher magnification of the cuticle composed of only a thin outer epicuticle (arrow) and an epicuticular filament layer (asterisk). ×30,000—scale: 0.5 μm.

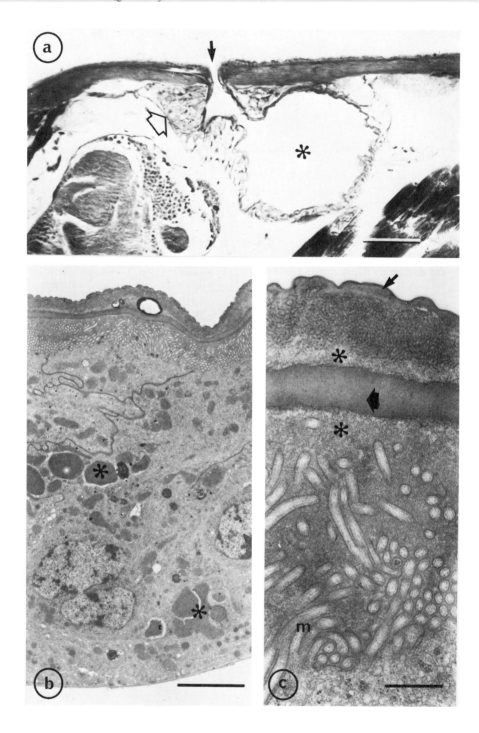

On histological sections, the frontal gland appears well developed (1,000 μm long and 150 μm wide, Figure 5.9c) and extends forward to the edge of the head. The glandular epithelium seems to be composed of only class 1 cells that are thicker (15 μm) on the gland floor. Under the cuticle of the triangular area, vacuolar class 3 units are present and are connected with the outside by the small openings surrounding the frontal pore.

Prorhinotermitinae

Due to the different evolutionary features of *Prorhinotermes*, Quennedey and Deligne (1975) separated this genus into a new subfamily (Prorhinotermitinae) within the family Rhinotermitidae.

In *Prorhinotermes simplex* the frontal pore opens into a small crater 50 μm in diameter, situated in the midpart of the head. The edge of the crater continues forward as two longitudinal cushions bordering a flared furrow that is completely smooth at the level of the clypeus.

The frontal gland is well developed in the form of a large thin pocket extending deep into the abdomen. The glandular epithelium (5–10 μm thick, Figure 5.9a) is composed only of class 1 cells that show a high degree of overlapping as indicated by the twisted pattern of the intercellular contact. The junctions are of the septate type, and localized in the upper part. On both of the other sides the plasma membrane is quite regular and shows neither microvilli on the upper side, nor invaginations along the basal lamina. The cytoplasm contains an oval-shaped nucleus (3–4 μm long) with peripheric aggregates of chromatin, long mitochondria (3 μm long) with a dense matrix, numerous free ribosomes, golgi bodies, and extensive smooth endoplasmic reticulum, and an abundance of microtubules fastened to the basal plasma membrane by hemidesmosomes. Large membrane-lined vesicles (possibly autophagic) with dense heterogenous

FIGURE 5.8. The frontal gland of *Reticulitermes santonensis*. (a) A sagittal section of the soldier head at the level of the glandular pore (small arrow) showing the anterior part of the frontal gland (asterisk) with a thin epithelium. Beneath the pore, the epithelium is thicker (large arrow). ×150—scale: 0.1 mm. (b) An electron micrograph of the gland composed of class 1 cells. Note the spots of electron-dense secretion in the cytoplasm (asterisks). ×6,200—scale: 3 μm. (c) A detailed view of the upper part of the cell with microvilli (m) and the cuticle composed of an inner layer (larger arrow) and the apical epicuticle (small arrow). Beneath both cuticular layers the extracellular spaces are filled with secretion (asterisks). ×32,000—scale: 0.5 μm.

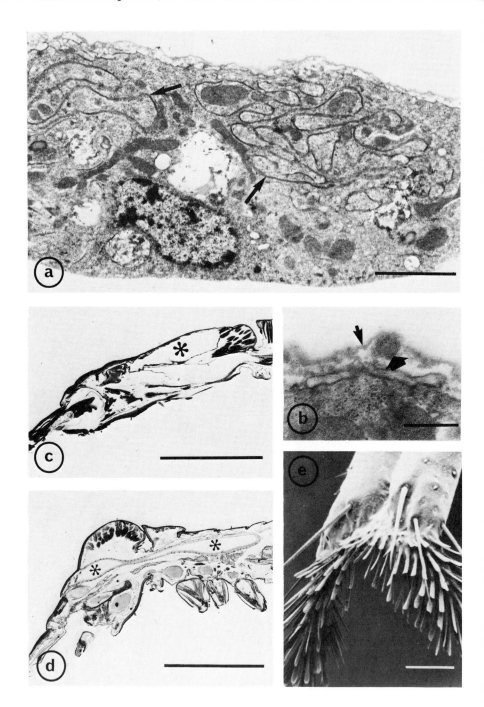

contents are also observed in the cytoplasm. The cell surface is covered by a slender, badly defined layer that is believed to be an outer epicuticle (Figure 5.9b).

The frontal secretion of *Prorhinotermes simplex* is toxic and is mainly composed of a nitro-compound: *trans* 1-nitro-1-pentadecene (Vrkoc and Ubik, 1974).

Rhinotermitinae

Within this subfamily the genera are more numerous and exhibit the most evolved frontal weaponery. This more advanced evolutionary state is shown in most cases by the presence of dimorphic soldiers whose minor nasutoid soldiers are characterized by reduced mandible size and an elongation of the clypeolabrum (Quennedy and Deligne, 1975).

Parrhinotermes may be considered the most primitive genus of this subfamily, since it possesses only a single type of soldier. In this, the frontal pore (20 μm in diameter) is at one end of a well-marked furrow that runs forward onto the frontoclypeus and the labrum. The labral end exhibits a daubing brush composed of hundreds of spatulated spines. In addition, long sensory bristles are present on the brush margin. The frontal gland, which appears less developed, is localized in a part of the head and is an ovoid sac (400 μm long and 200 μm wide) whose lower epithelium is thick (10 μm) and seems to have only a secretory function.

Schedorhinotermes possesses dimorphic soldiers. The frontal weapon of the major is similar to *Parrhinotermes*, except it is larger and has a less bushy apical brush. A lengthening of the labrum occurs in the minor soldier, causing it to overlap the larger part of the mandibles whose acute apex rubs against the well-spined daubing brush. In addition, several types of sensory organs are present in these cephalic

---

FIGURE 5.9.   The frontal weapon of Rhinotermitidae soldiers. (a) An electron micrograph of a *Prorhinotermes simplex* frontal gland. The epithelium is composed of overlapping class 1 cells as indicated by the twisted pattern of intercellular contacts (arrows). ×11,000—scale: 2 μm. (b) Higher magnification of the cell surface covered by a badly defined thin layer considered to be an outer epicuticle (small arrow). Beneath it a droplet of secretion (large arrow) is seen. ×30,000—scale: 0.5 μm. (c) A sagittal section of a *Termitogeton umbilicatus* head, the frontal gland (asterisk) is located in the head. ×28—scale: 1 mm. (d) In *Dolichorhinotermes longilabius* minor soldier, the frontal gland (asterisks) extends into the abdomen. ×26—scale: 1 mm. (e) The daubing brush of *Rhinotermes marginalis* minor soldier. ×430—scale: 30 μm. (From Quennedey and Deligne, 1975).

areas. Based on its ultrastructure alone a functional interpretation of the *Schedorhinotermes putorius* minor soldier frontal weapon has been proposed (Quennedey, 1975b): when the soldier is disturbed it takes a waiting posture, mandibles open, and a drop of secretion flows out of the frontal pore and coats the labrum. Short chemo- and mechano-sensory bristles, scattered along the median furrow, inform the soldier of the direction and the quantity of the flow. The soldier then brings its labral brush into contact with the enemy's body by lifting the labrum. At this time, certain long chemo- and mechanosensory bristles located at the brush margin inform the termite about the contact and its chemical nature, while campaniform sensillae provide information about the distortion of the labral cuticle during the brushing. In contrast to the major soldier, the mandibles of the minor soldier are not used often (Quennedey et al., 1973).

The chemical nature of *Schedorhinotermes* frontal gland secretion is well established. Several alkanones, alkenones, and alkadienones have been identified and are known to have a toxic effect (Quennedey *et al.*, 1973; Prestwich, 1975; Prestwich *et al.*, 1975).

In both soldiers the frontal gland is highly developed and has the same organization. It takes up a great part of the soldiers' body and has a thin glandular wall (10 μm thick) composed of only class 1 cells (Quennedey, 1973, 1975c, 1978). In the minor soldier the glandular epithelium, a short distance (80 μm) from the frontal pore, where it begins, has high epithelial cells that reduce the glandular aperture.

The class 1 cells found in the gland (Figure 5.10b) contain a basal nucleus with a denser nucleolus and numerous chromatin aggregates, mitochondria, microtubules, and an abundance of glycogen granules often associated with the large myeloid vesicles of secretion. The apical border of the plasma membrane is composed of tightly packed short (2 μm long) microvilli that have inner microfilaments. On the hemocoel side, short membranous invaginations are seen and the lateral contorted intercellular space is apically reinforced by a septate junction.

The structure of the cuticle and the aspect of the secretion stored under it vary along the length of the gland. Near the aperture, where the cells are highest, their cytoplasm contains more microtubules but no secretory material. The cuticle is composed of a thick (2 μm) mesocuticle and an apical epicuticle (Figure 5.10a). The remainder of the gland contains apparent modifications. Within the head, small cavities exist in the mesocuticle, while an electron-dense secretion collects below it (Figure 5.10c). The cavities are twisted in an arrangement similar to the orientation of the endocuticular lamellae. Histologically, these stain in a blue color similar to that of the endocuticle with Heidenhain's Azan, but no other evidence exists to suggest that they are the same. A bilayered (inner and outer) epicuticle is clearly visible above the mesocuticle. Beneath the cuticle the layer of secretion (1 μm thick) is easily recognizable and is a result of the aggregation of small secretion droplets. These extracellular structures gradually change within the abdomen until the cuticle appears composed of a thin epicuticle with a loose inner layer (Figure 5.10c). Beneath this secretion form thick (2 μm) hexagonal plates of a dense material close to the cell surface and corresponding to the boundary of each class 1 cell (Figure 5.10d and e). In the end part of the gland, it is no longer in contact with the epidermal surface; it is possible, however, that this feature of the gland could be an artifact.

Among the other Rhinotermitinae, two genera, *Rhinotermes* and *Dolichorhinotermes*, possess dimorphic soldiers, and the genus *Acorhinotermes* possesses a single minor soldier. The major soldiers, like those of *Schedorhinotermes*, have well-developed mandibles, which are strongly denticulated and powerful. In *Dolichorhinotermes*, the size of the soldier is reduced and the labrum, although elongated, is only two-thirds of the length of the mandibles. These mandibles hardly seem able to grasp the enemy. The minor soldiers of the three genera are called nasutoid soldiers because of the structural modification of their heads. They have in common vestigial and nonfunctional mandibles, and a bushy brush handle at the end of an elongated indivis rostrum composed of unrecognizable frontal, clypeal, and labral regions (Figure 5.9e). In *Rhinotermes* and *Dolichorhinotermes* the frontal pore opens at the base of the rostrum, where the furrow is situated. In *Acorhinotermes*, the frontal pore opens further forward and behind it the rostrum is tube shaped. In all three cases the nasutoid structure permits the quick daubing of the opponent with the frontal secretion (Quennedey and Deligne, 1975; Quennedey, 1975a).

Important differences exist between the frontal glands of minor and major soldiers. Sannasi (1969) has described a very small glandular invagination (50 μm in diameter) in the *Rhinotermes magnificus* major soldier. It appears to be composed of different types of cells in a trilayered arrangement. We have also found a small gland in the *R. marginalis* major soldier, although unfortunately the material was badly preserved and did not permit description of the glandular organization. In contrast, the frontal gland of minor soldiers is well

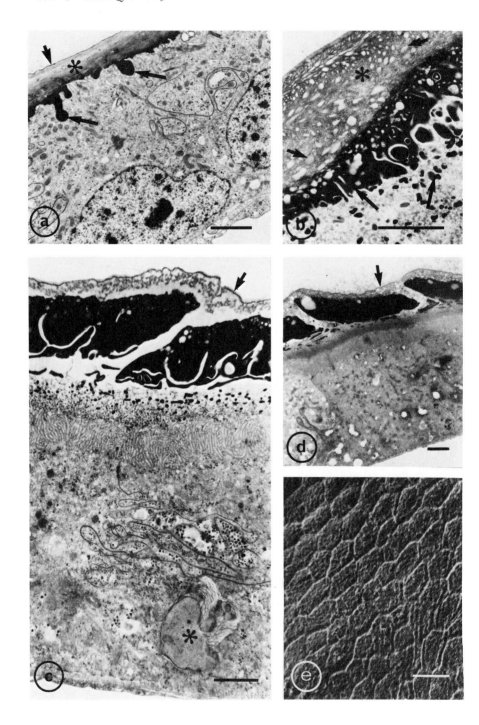

developed. In the *Dolichorhinotermes longilabius* and *Acorhinotermes sub-fusciceps* minor soldiers, the frontal glands extend as far as the anterior third of the abdomen (Figure 5.9d).

Termitidae

The populous family Termitidae has been taxonomically redescribed and is now composed of four well-separated subfamilies: the Macrotermitinae, the Apicotermitinae, the Termitinae, and the Nasutitermitinae (Sands, 1972). Knowledge of the various subfamilies differs greatly, and that of numerous genera is very limited. Due to this situation we discuss only a few genera, which may be considered examples only and cannot reflect the evolution of the whole family.

Macrotermitinae

Twelve genera are included in this subfamily, and scanning electron microscope data are available (Quennedey, 1978). In the six species *Allodontotermes giffardii*, *Protermes minutus*, *Odontotermes smeathmani*, *Hypotermes xenotermitis*, *Microtermes congoensis*, and *Synacanthotermes sp.* neither frontal aperture nor frontal differentiation is visible. In five other genera, *Sphaerotermes*, *Ancistrotermes*, *Macrotermes*, *Acanthotermes* and *Pseudacanthotermes*, a frontal pore is present; the first three have a very narrow aperture while the remaining two exhibit a very unusual frontal appendage.

The existence of a frontal pore is caused by the invagination of the frontal gland; however, this does not mean that the lack of such an aperture can be interpreted as an absence of the gland. Light microscopy of longitudinal sections of the soldier heads of *Allodonto-*

FIGURE 5.10. The structural variations in the cuticle of *Schedorhino-termes putorius* minor soldier. (a) Near the frontal aperture the cuticle is thick and composed of epicuticle (small arrow) and mesocuticle (asterisk). Beneath it secretory droplets (large arrows) are seen. ×2,200—scale: 5 μm. (b) The cuticle (asterisk) in the head has twisted cavities (small arrows) and dark secretory droplets (large arrows) amass beneath it. ×9,200—scale: 2 μm. (c and d) In the thorax (c) and a great part of the abdomen (d) the secretion forms thick plates of dense material. The cuticle consists only of a thin epicuticle (arrows). Each plate is free and corresponds to the boundary of the under cell. Note also the myeloid secretion (asterisk) into the cytoplasm. (c) ×5,500; (d) ×3,000—scales: 2 μm. (e) A scanning electron micrograph of the glandular cuticle with its hexagonal pattern corresponding to the boundaries of the cells. ×1,000—scale: 10 μm.

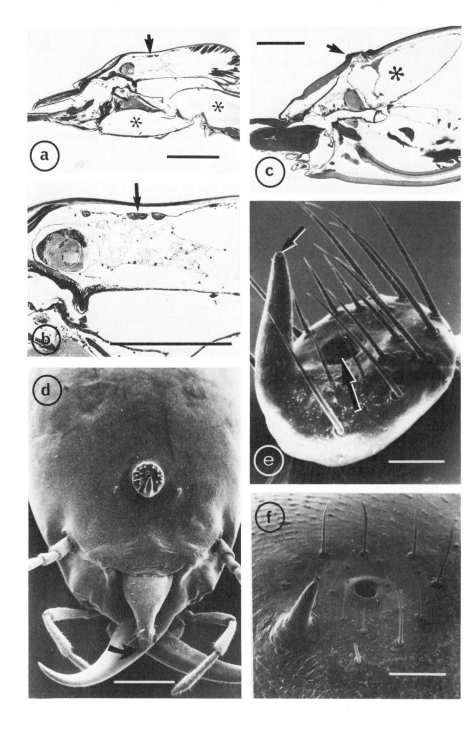

termes, Hypotermes, and *Microtermes* shows that these genera lack the frontal gland. In them the cephalic epithelium at the frontal level appears very thin without any thickening or infolding. On the other hand, *Odontotermes* has a few large cells (20 μm long) scattered in the frontal epithelium in an area 400 μm in length at the back of the brain (Figure 5.11a and b) and the question arises as to whether such cells may be considered to be a remnant of the frontal gland. Meanwhile, these cells cannot be confused with the thickening generally observed under the fontanel of workers and alates due to the attachment of the tentorial–fontanellar muscles. These muscles demarcating the lateral extent of the fontanel are observed sagitally on both sides of the area of large cells.

In genera in which the frontal gland is developed it is always present within the head of the soldier (Figure 11c, 12a). In the monomorphic soldiers of *Sphaerotermes sphaerothorax,* the frontal pore (15 μm in diameter) opens at the middle of the hairy head, but no data are available concerning the size of the gland. In the dimorphic soldiers of *Ancistrotermes latinotus,* the frontal pore has a rather rectangular shape (5 × 7 μm) and is situated at the apex of a small hump on the head; here also the size of the frontal gland is unknown.

In the case of the dimorphic soldiers of *Macrotermes bellicosus* (= *Bellicositermes natalensis*) the frontal gland has a round opening (20 μm in diameter) and is also situated on a small hump at the limit of the third anterior section of the head (Figure 5.12b and c). The glandular epithelium is more developed in the major soldier. It appears as a pear-shaped, thin-walled (20 μm thick) organ located at the back of

---

FIGURE 5.11.   The frontal weapon of Macrotermitinae soldiers. (a and b) The frontal gland is absent in *Odontotermes*, nevertheless a few large cells are observed in the frontal epithelium (arrows). Note the great volume of the salivary reservoirs (asterisks). (a) ×23; (b) ×75—scales: 0.5 mm. (c) A sagittal section of *Acanthotermes* medium soldier head showing the great development of the frontal gland (asterisk) and the base of the turret (arrow). ×26—scale: 0.5 mm. (d and e) Scanning electron micrographs of *Acanthotermes* medium (d) and minor (e) soldiers showing their peculiar frontal apparatus. It is composed of a turret topped by a circular plate in the middle of which is seen the opening of a primary frontal pore (large arrow) surrounded by bristles. At the apex of the anterior horn a secondary minute frontal pore (small arrow) is also seen. Note the hyaline tip at the end of the labrum (curved arrow). (d) ×32—scale: 0.5 mm; (e) ×300—scale: 50 μm. (f) In *Pseudacanthotermes* minor soldier, the turret is lacking but the pores, the horn, and the bristles are present. ×150—scale: 100 μm.

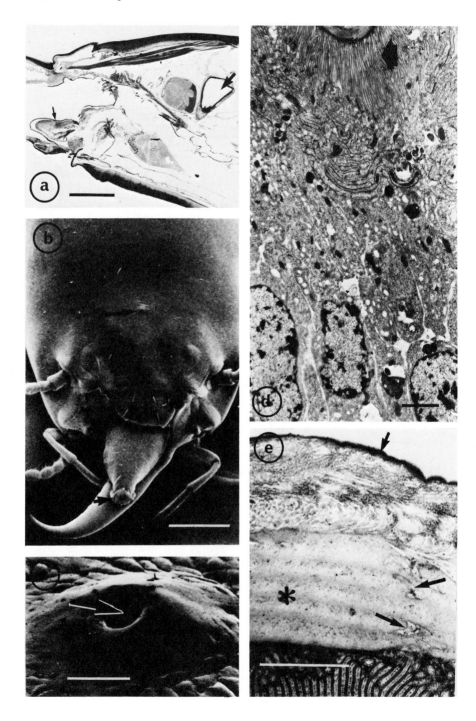

the brain with a vast lumen into which the secretion accumulates (Figure 5.12a). In contrast, in the minor soldier, the reservoir is extremely reduced and the gland is small (100 μm in diameter) and is mainly composed of a thick (50 μm) epithelium. These differences in the size of the glands in the two types of soldier have been previously reported by Cmelik (1971) for *M. goliath*, in which the gland of the major soldier fills most of the head cavity (3 mm long and 2 mm large) and by Prestwich *et al.* (1977) for *M. subhyalinus*, where the glandular reservoir of the major soldier is 500 times larger than that of the minor one. Preliminary ultrastructural data show that the frontal gland of the major soldier of *M. bellicosus* is composed of class 1 cells (Figure 5.12d). Its main structural feature is the numerous invaginations that exist in the plasma membrane facing the hemocoel. The apical plasma membrane forms numerous long microvilli and the contorted intercellular space shows a septate junction along its apical part. In the cell cytoplasm the nucleus is located basaly; mitochondria, endoplasmic reticulum, and microtubules are abundant. The glandular cuticle lining the reservoir is composed of a lamellate endocuticule (3 μm thick) crossed by twisted pore canals and a thin bilayered epicuticle under which epicuticular filaments are seen (Figure 5.12e).

The chemical nature of the frontal secretion is known for only two species of *Macrotermes*. Cmelik (1971) has isolated *n*-alkanes and isoalkanes mixed with fatty acids, sterols, and phospholipids from *M. goliath*. Prestwich *et al.* (1977) have found saturated and unsaturated hydrocarbons from *M. subhyalinus*. In this species the secretion has a delayed toxic effect when it penetrates into the wound inflicted by the mandibles.

FIGURE 5.12.   The frontal weapon of *Macrotermes bellicosus* soldiers. (a) In the major soldier the frontal gland located at the back of the brain has a vast lumen (large arrow). Note the hypopharyngeal part of the cibarial gland (small arrow). ×25—scale: 500 μm. (b) A scanning electron micrograph of a minor soldier head, the arrows give the locations of the frontal pore (large arrow) and the hyaline tip (small arrow). ×17—scale: 1 mm. (c) A high-magnification micrograph showing the minute pore (arrow) situated at the apex of a small hump. ×430—scale: 20 μm. (d) A transmission electron micrograph of the major soldier gland composed of class 1 cells. Note the large number of long microvilli (arrow) under the cuticle. ×5,000—scale: 2 μm. (e) A high-magnification micrograph of the cuticle, composed of a lamellate endocuticle (asterisk) crossed by twisted pore canals (large arrows), and of an epicuticle (small arrow). ×24,000—scale: 1 μm.

The frontal weapon of the trimorphic soldiers of *Acanthotermes acanthothorax* is among the strangest and the most questionable specializations observed in the soldiers of termites (Figure 5.11d). It is in the minor soldier that such development appears most sophisticated. In appearance it resembles a turret composed of a basal foot holding a circular plate (200 μm in diameter and 50 μm thick), in the middle of which the primary frontal pore (30 μm in diameter) opens. About 20 long, articulated bristles are arranged in two concentric circles around the frontal pore. A small curved horn (100 μm) with a minute apical opening (2 μm) stands at the anterior of the plate (Figure 5.11e). The same structure was found in the two other soldiers, but in the medium soldier the inner foot is absent and the plate adjoins the head capsule. This phenomenon is more pronounced in the major soldier, where the circular plate is partly apparent. The frontal gland is well developed and extends backward into the head cavity. In the medium soldier (Figure 5.11c) the glandular pouch is 1200 μm long, 700 μm wide, and 450 μm thick and is composed of a thin epithelial wall (10 μm thick). The gland of the minor soldier is smaller in size (500 × 300 × 200 μm) but the glandular epithelium is twice as thick. Below the turret there is a cluster of other cells that seem related to the hollow anterior horn.

The dimorphic soldiers of *Pseudacanthotermes spiniger* have a frontal apparatus similar to that of *Acanthotermes* major soldiers, and the turret is lacking in both soldiers. The large frontal pore is surrounded by a double circle of sensory bristles and an anterior horn on which a small apical opening is found (Figure 5.11f).

Apicotermitinae

Sands (1972) has classified the 35 genera of this subfamily into two branches: the *Apicotermes* branch where soldiers exist, and the *Anoplotermes* branch which contains the soldierless termites. Very little is known about the soldiers in the Apicotermitinae, but some data from observations performed on two genera of this subfamily do exist. The soldiers of *Foraminitermes sp.* have a small frontal pore (7 μm in diameter) situated on top of a hemispheric bulge (0.4 mm high), which is surrounded by fifty small sensory bristles. In *Rostrotermes cornutus* the frontal pore is marked by a little dip at the back of a frontal prominence into which numerous epidermal class 3 units open. In both genera the exact role of the frontal gland in the soldier defense mechanism remains to be studied thoroughly.

Termitinae

According to the reclassification by Sands (1972), this subfamily contains three evolutive branches. The first is the *Microcerotermes* branch, included previously in the Amitermitinae subfamily (whose other members were dispatched into the Apicotermitinae). The other Termitinae genera belong to the *Thoracotermes* and the *Termes* branches.

Within the *Microcerotermes* branch three genera have been studied by scanning electron microscopy: *Microcerotermes, Cephalotermes*, and *Pseudhamitermes*. The frontal pore is absent in *Microcerotermes fuscotibialis* and *M. parvus* soldiers, but in the midpart of the frontal area there is a transverse furrow. It is not known whether this furrow is due only to muscle attachments or if it is related to a thickening of the frontal epithelium. On the contrary, an oval frontal pore (20 μm long) is clearly seen at the limit of the anterior third of the elongated hairy head of the *Cephalotermes rectangularis* soldier. In the neighboring genera *Amitermes evuncifer, A. obeuntis*, and *Pseudhamitermes khmerensis*, the frontal weapon has a more elaborate structure. The frontal pore (30 μm in diameter) has a vertical opening protected by a frontal fold in the case of *Pseudhamitermes*. The pore is concentrically surrounded by 20 long sensory bristles. In front of it a median gutter (poorly defined) extends to the frons and the bilobated clypeus. The frontal gland is well developed and located within the head. The chemical nature of the frontal secretion varies greatly from one species to another. Wadhams *et al.* (1974) have isolated a tricyclic ether, 4,11-epoxy-*cis*-eudesmane, in *Amitermes evuncifer*, which performs a formicidal activity. In five other species belonging to the same genus, such varying products as methylketones, alcohols, and sesquiterpene hydrocarbons have been found (Prestwich, 1975; Meinwald *et al.*, 1978).

The soldiers of the *Thoracotermes* and the *Termes* branches show many different morphological variations of the head capsule. The frontal weapon of *Cubitermes fungifaber* (Quennedey, 1978) is similar, with minor variations, to that of many of its neighboring genera. Nevertheless, we cannot use this example to give a general idea of the frontal weapon evolution within these two branches of Termitinae, where the mandibular weapon is very sophisticated in the more evolved genera (Deligne, 1965a, b, 1970, 1971; Deligne *et al.*, 1981). Numerous cases of frontal gland regression also exist in those two branches.

In *C. fungifaber* the head capsule has a rectangular shape and the

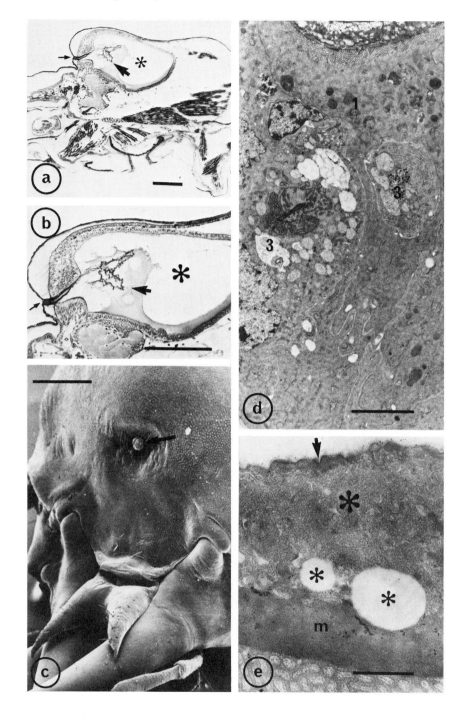

frons is convex and very obvious (Figure 5.13a and c). The vertical frontal pore (30 μm in diameter) opens beneath a small protuberance. (In *Termes* and the closely related "snapping" genera, this protuberance develops into a sharp-pointed extension.) The frontal pore is easily located due to a tuft of many sensory bristles surrounding it. These bristles are long (150 μm), longitudinally grooved, articulated, and bent toward the frontal pore. Some are both chemo- and mechanosensory, while the others are only mechanosensory. Small openings of epidermal class 3 units are also seen between the base of the bristles. Due to the convex nature of the frons the frontal pore overhangs the clypeus and the labrum. The latter is short, indented, and moves back close to the clypeus during the discharge of the frontal secretion (Deligne, 1970). Thus the secretion crosses the tuft of bristles, flows down the clypeus, and onto each side of the labrum before permeating the sword-shaped mandibles (Figure 5.13c).

The frontal gland is well developed (500 μm long with a 0.2-mm^3 capacity) and is located in the anterior half of the head (Quennedey, 1973, 1978). The glandular epithelium (20 μm high) is thicker (60 μm) near the glandular aperture. In the gland lumen a heterogenous secretion is preserved by histological fixatives (Figure 5.13b). The glandular epithelium is composed of mixed class 1 and class 3 units (Figure 5.13d). The class 1 cells form tightly packed and numerous microvilli beneath the glandular cuticle, while the basal plasma membrane has short invaginations. The cytoplasm appears electron dense because of the large number of microtubules and vesicles of smooth endoplasmic reticulum. It also contains an oval-shaped nucleus, mitochondria, free ribosomes, and abundant dark lysosomes. The class 3 units are scattered between the class 1 cells. The basal secretory cell is oval shaped and possesses a large basal nucleus. The cytoplasm is characterized by well-developed golgi bodies secreting numerous large clear vesicles with a granular content. The same

---

FIGURE 5.13. The frontal weapon of *Cubitermes fungifaber* soldier. (a and b) A sagittal section of the head. The frontal gland (asterisks) opens forward (small arrows) and a heterogenous secretion is present within the lumen (large arrows). (a) ×17; (b) ×35—scales: 0.5 mm. (c) In this scanning electron micrograph a droplet of secretion is seen going out of the pore (arrow) surrounded by a tuft of bristles. ×80—scale: 200 μm. (d) A low-magnification electron micrograph of the frontal epithelium composed of class 1 cells (1) and class 3 units (3). ×4,200—scale: 4 μm. (e) A high-magnification micrograph of the cuticle. Secretory droplets (small asterisks) are observed between the mesocuticle (m) and the pitted epicuticle (arrow) overlapping the epicuticular filament layer (large asterisk). ×40,000—scale: 0.4 μm.

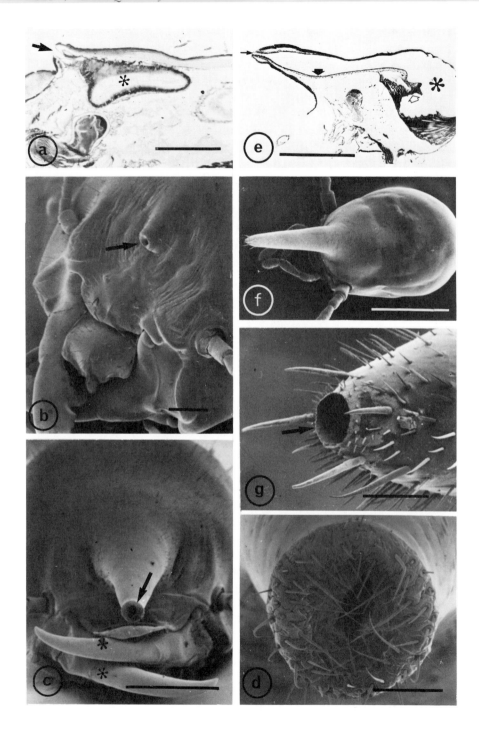

glandular material is seen in the central extracellular space of the cell bordered by the microvilli. It is also found in the lumen of the duct of the canal cell and in the glandular reservoir. The cuticle lining of the reservoir is composed of an apical bilayered epicuticle and an inner mesocuticle between which there is an epicuticular filament zone where numerous clear vesicles secreted by the class 1 cells collect (Figure 5.13e). Afterward this secretion is mixed with the products of the class 3 units in the glandular reservoir where it forms a hetero-genous mixture of small clear vesicles bound to a dark-phase material.

Four major diterpene hydrocarbons have been recently isolated from several species of *Cubitermes*: cubitene, two isomers of a cembrenoid 14-membered ring and neocambrene-A (Meinwald *et al.*, 1978; Prestwich *et al.*, 1978). The cubitene 1 had been previously isolated from the frontal secretion of *C. fungifaber* but was not characterized (Quennedey, 1973).

Nasutitermitinae

The frontal weapon is always present in the soldiers belonging to the large subfamily Nasutitermitinae.

In the most primitive genus, *Syntermes*, the enormous soldiers (compared in size to the other Nasutitermitinae soldiers) possess well-developed mandibles and a frontal gland opening at the apex of a short tube (Figure 5.14a, b). Two evolutionary branches are then distinguished: the *Procornitermes* and the *Paracornitermes* branches. Within each of these branches the less evolved genera have functional mandibles and a lengthened frontal tube (Figure 5.14c). Thus, these

---

FIGURE 5.14. The frontal weapon of Nasutitermitinae soldiers. (a) A sagittal section of *Syntermes* head with the frontal gland (asterisk) opening at the apex of a short tube (arrow). ×35—scale: 0.5 mm. (b) A frontal scanning view showing the short frontal tube (arrow). ×20—scale: 0.5 mm. (c) *Cornitermes* soldiers have a nasus (arrow) and well-developed mandibles (asterisks). ×50—scale: 0.5 mm. (d) A detailed view of the nasus ending forming a snout on which about 100 sensory bristles are seen. ×350—scale: 50 μm. (e) A sagittal section of *Trinervitermes* major soldier with the retort-shaped gland (asterisk). The class 1 cells are situated in the posterior part (white arrow) of the gland surrounded by a powerful mandibular neck (black arrow). The small black arrow indicates the frontal opening. ×40—scale: 50 μm. (f) A low-magnification micrograph of a typical nasute soldier head (*Trinervitermes geminatus* major soldier) where mandibles are lacking. ×40—scale: 50 μm. (g) A high-magnification view of the end of the nasus; the glandular pore (arrow) is flanked by four tall mechanosensory bristles. ×350—scale: 50 μm.

form a transition state between the well-known nasute soldier form characterized by the pear-shaped head capsule and atrophied mandibles (Figure 5.14f) and the primitive *Syntermes*. The unusual shape of the head of the nasute soldiers has stimulated a great amount of research, and more is now known of the evolutionary trends in the Nasutitermitinae than of any other termite group.

The following is a detailed description of the frontal weapon of an evolved species, *Trinervitermes geminatus*, which possesses dimorphic nasute soldiers (Quennedey, 1973, 1978). The entire head is 2.2 mm in length in the major soldier and 1.5 mm in the minor. The rostrum or nasus overhanging the front of the head is 0.8 mm in the major and 0.6 mm in the minor soldiers, and varies in width from 0.2 mm at the base to 0.05 mm at the apex. The circular frontal pore is also a little larger in the major soldier (30 μm in diameter against 25 μm; Figure 5.14g). Numerous more or less short sensory bristles are seen on the surface of the apical third of the rostrum and the frontal pore is flanked by short bristles and four tall bristles (80 μm long). All these bristles are mechanosensory and have a single neuron ending at their base.

Numerous class 3 unit openings are scattered over the entire surface of the nasus. They have the same ultrastructural features as the class 3 units of the frontal gland described below. We take into account the distribution of the cells of the rostrum separately because they show a considerable variation from one genus to another.

We give additional unpublished scanning observations about a few species belonging to the *Procornitermes* branch. In the mandibulate *Cornitermes* genus the nearly bristleless nasus has a membranous and circular snout on which a hundred long sensory bristles bend toward the frontal pore (Figure 5.14d). In the evolved genera studied (*Longipeditermes, Nasutitermes, Tumulitermes, Bulbitermes*, and *Hospitalitermes*) the bristle distribution is approximately the same as *Trinervitermes* nasus.

Most of the Nasutitermitinae have a retort-shaped frontal gland (Figure 5.14e) located at the back of the head, which has a slender neck that passes above the brain before entering the nasus (Holmgren, 1909; Bonneville, 1936; Noirot, 1969; Quennedey, 1973, 1978). *Trinervitermes* is similar to *Nasutitermes* (Bonneville, 1936) as the glandular epithelium is composed of two classes of cells. The class 1 cells form the glandular epithelium (40 μm thick) in the basal part of the retort while the class 3 units lie along the neck. In the upper part of the retort the glandular wall is thinner and without secretory cells. Despite the difference in size of the two soldiers the frontal gland has the same structural organization, the only difference being that it is three times larger (0.3 mm³) in the major soldier.

The class 1 cells of the gland (Figure 5.15a) have a prismatic shape (40 μm long and 10 μm wide). On the apical side the brush border is formed with very long (10 μm), tightly packed and bent microvilli with inner microfilaments. Numerous dark spots of secretion (about 0.5 μm large) fill up the extracellular space between the microvilli. In the cytoplasm, the oval-shaped nucleus (7 μm long) is located in the basal half of the cell and microtubules are abundant. Large, mainly rounded, dense vesicles are often observed. The intercellular border is quite regular except in the apical part, where a septate junction shows a zigzag pattern. Along the basal plasma membrane, overlapped by a thick (2 μm) basal lamina, the hemidesmosomes are particularly abundant.

Above the class 1 cells the cuticle (Figure 5.15b) is up to 3 to 5 μm thick and is composed of a very thin outer epicuticle (20 nm) pitted by the epicuticular pores. Below it there is a layer (1.5 μm) composed of epicuticular filaments with small (0.3 μm in diameter) clear vesicles positioned intermittently. The deepest layer of the cuticle is made of endocuticle, in which three laminae are recognizable and pore canals are lacking. Between it and the epicuticular filament layer there is a small and irregular space full of a granular secretory material.

The class 3 units do not form a distinct glandular area but are scattered along the neck of the gland. At this point the cuticular duct is surrounded by flat epithelial cells and the class 3 units project out of them. The cytoplasm of the secretory cell has a vacuolar appearance because of the abundance of large vesicles (Figure 5.15c). The vesicles have a granular content and originate from the golgi bodies, which are highly developed. Glycogen and free ribosomes are also numerous. The granular content of the vesicles is discharged by exocytosis into the inner extracellular space, which is lined by long microvilli. No junction is seen in the intercellular space between two class 3 units, but a septate junction is always present between the secretory cell and its fellow canal cell. Each canal cell has a denser cytoplasm, and for a short distance these cells surround the cuticular duct. This duct is composed of a bilayered epicuticle and has no adherent cells from the point where it crosses the glandular cuticle of the neck. This cuticle looks ringlike in cross section and is composed of a thick (6 μm) exocuticle and an epicuticle pitted by epicuticular pores. Around the central ring of the neck there is another cuticular ring in which the cuticular layers are in reverse order to those of the glandular neck

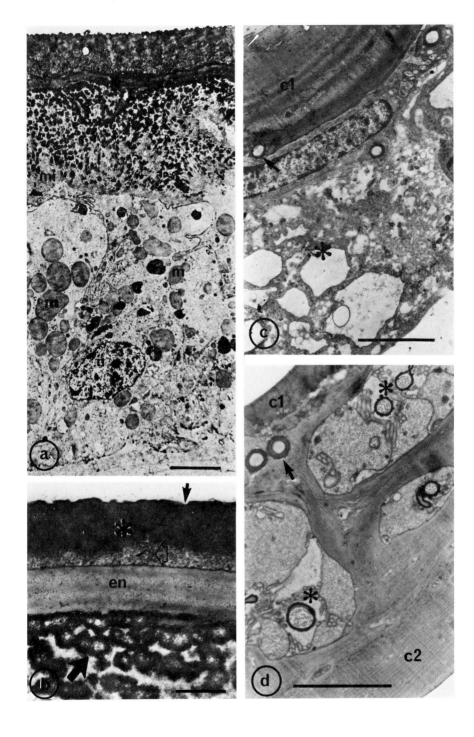

(Figure 5.15d). This very thick cuticle is crossed by the cuticular ducts of the class 3 epidermal units and by the dendritic ending of the numerous mechanosensory bristles.

In *Trinervitermes* the class 3 unit openings are mainly located on the anterior part of the nasus. In other genera belonging to both branches of the subfamily they may be more numerous and be more widely distributed about the nasus. Thus, in *Eutermellus* a hundred of such cells open onto the bottom of oval-shaped hollows (Deligne, 1973). In *Verrucositermes* (Deligne, 1981) the cells are clustered into small groups and open onto the faces of numerous polyhedric tubercules scattered at the basal half of the nasus and at the anterior part of the head. Such features have lead Deligne (1973, 1981) to consider it a special glandular area called the "rostral system." An analogous glandular system is also present in other termites not belonging to Nasutitermitinae. Numerous epidermal cell openings are seen on the anterior part of the head capsule of two Apicotermitin soldiers, *Rostrotermes* and *Machadotermes*. It is probable that such openings may be found in other species, but their exact role—particularly in defense—remains in doubt.

For a long time the only biochemical data on the secretion of nasute soldiers were those of Moore (1964, 1968). He demonstrated the presence of volatile terpenoids that act as a solvent for a resinous material. This viscous gluelike secretion was ejected a short distance

---

FIGURE 5.15. The frontal gland of *Trinervitermes geminatus* major soldier. (a) A general view of class 1 cells present in the posterior part of the gland. Note the rounded mitochondria (m) in the cytoplasm, the apical tightly packed microvilli (mi) with abundant dark spots of secretion. ×2,500—scale: 5 μm. (b) A detailed view of the cuticle overlapping the class 1 cells. Beneath the pitted epicuticle (small arrow), a layer of epicuticular filaments (asterisk), an extracellular space filled with granular material (large arrow), and three endocuticular laminae (en) are successively seen. Note the numerous dark spots of secretion beneath the cuticle (large arrow). ×13,000—scale: 1 μm. (c) The class 3 units are scattered along the neck of the gland. The cuticular ring lining the glandular lumen is particularly thick (c1) and is crossed by the canal cell ducts (arrow). The secretory cell of each unit (asterisk) has numerous vesicles, giving it a vacuolar appearance. ×8,000—scale: 2 μm. (d) A cross section of the nasus ending, where the two concentric cuticular rings, the glandular cuticle of the neck (c1), and the nasus cuticle (c2) interlock. The inner glandular cuticle (c1) has its cuticular layers reversed with respect to the nasus cuticle (c2). Between them the duct of a canal cell (arrow) and the dendritic endings of the mechanosensory bristles (asterisks) are seen. ×13,000—scale: 2 μm.

by the soldier, thus entangling the opponent. Since then a wide range of monoterpene hydrocarbons—the "solvents"—have been isolated (see references in Prestwich, 1979; Deligne *et al.*, 1981) and the chemical nature of the "resinous" material, composed of several diterpenes, has been well established (Prestwich, 1979; Prestwich *et al.*, 1979; Braekman *et al.*, 1980).

It is interesting to attempt to correlate the two classes of chemical products with the two types of cells observed in the frontal gland. It is possible that the class 3 units could secrete the "solvent" part of the secretion, thus preventing the plugging of the neck by the diterpenes. In *Trinervitermes* the "rostral system" could play a similar role, that is, preventing the adhering of the secretion onto the rostrum. In my opinion both class 3 units may be considered as the same glandular area, discharging both into the neck and directly out onto the nasus. However, only biochemical data can verify such a hypothesis.

DIVERSITY OF SOLDIER DEFENSE GLANDS

We have only considered the soldier caste whose defensive role in the termite society is well established and has been known for a long time.

The case of the epidermal glands associated with the mandibles of *Machadotermes* soldiers appears to be a peculiar defensive specialization; nevertheless, this epidermal system can bring nearer the numerous epidermal cells frequently observed at the surface of the anterior part of the head and of the cephalic appendages of numerous species of soldiers. Meanwhile, the exact function of such epidermal cells remains unknown.

The cibarial glands seem to be only differentiated in the genera whose salivary glands have a defensive function. Their location around the cibarium, into which the salivary fluid passes, tends to favor the idea of a complementary effect of both glands' secretion. It would be interesting to resolve whether the existence of a hyaline tip in the soldiers of the more primitive genera of Nasutitermitinae is correlative with a possible development of the cibarial glands. But in these latter genera, the soldiers do not possess a salivary defensive system.

As for the hypertrophied salivary glands of some soldiers, no major modification occurs in their morphology with regard to the workers' glands, except for the great development of the reservoirs. Such a defensive system is observed in the single genus belonging to the most primitive family (Mastotermitidae) and in genera belonging to the most evolved: Termitidae (Macrotermitinae, Termitinae: *Globitermes*). So such a specialization is probably due to a convergence or arises from a parallel evolution among different families.

The frontal gland appears as a peculiar organ without equivalent in other insects. Such a gland, undifferentiated in lower termite families, suddenly develops in Rhinotermitidae and Termitidae, particularly in the soldier caste. Little is known about the structure, distribution, or role of the frontal gland in workers and imagoes; nevertheless a defensive function seems doubtful. In both frontal and salivary glands, the great development of the storage areas is noteworthy. Thus, a great amount of secretion can be stored and suddenly discharged; the secretion is released either by contraction of the mandibular adductors (in nasute soldiers) or by pressure of the entire body squeezing the glandular reservoir. Therefore, due to its muscular strength and the small diameter of the frontal opening, the nasute soldiers can project the secretion to a distance of many centimeters.

We do not recapitulate here the various steps of the evolution of the soldiers' frontal weapon; again we invite the reader to refer to articles dealing with the subject (Quennedey and Deligne, 1975; Deligne *et al.*, 1981). Briefly, a frontal pore seems to be always present when a frontal gland exists. In the majority of cases a single opening is observed, often with an anterior migration of the frontal pore (Coptotermitinae, Rhinotermitinae, and Nasutitermitinae). In some cases, very strange and highly sophisticated adaptations appear (Rhinotermitinae, some Macrotermitinae). Therefore, such a weapon may be considered the most developed defensive system in termite soldiers.

Generally a single glandular system was encountered in the soldiers, yet some Macrotermitinae soldiers possess both types of defensive glands. The salivary glands are always developed, while the frontal gland is small in three genera and well developed in only two (*Acanthotermes* and *Pseudacanthotermes*). The salivary glands may be considered as the basic and preponderant defensive system of the subfamily and the frontal gland appears as a supplementary development that may be present in evolved genera. Nevertheless, this gland is not present in *Synacanthotermes*, a genus closely related to *Acanthotermes*.

CYTOLOGICAL FEATURES OF THE FRONTAL GLAND

We now consider the cytological and structural aspects of the evolution of the frontal gland. The distribution of the different classes of cells constituting a gland appears to be a good criterion for following the evolutionary steps of sternal and tergal glands of termites (Ampion and Quennedey, 1980). With regard to the frontal gland, the class 1 cells may be found alone (*Psammotermes, Reticulitermes, Prorhinotermes, Schedorhinotermes*, and *Macrotermes*) or associated with the class 3 units (*Coptotermes, Cubitermes*, and *Trinervitermes*) apparently without regard to phylogenetic relationships. This fact favors a diverse and free evolution among each family. Because of the great diversity of the chemical products isolated from the frontal secretions of termites, it is no wonder that we find such diversity in the glandular structures.

Alternatively, it is difficult to draw a parallel between the chemical nature of the secretion and the nature of the secretory droplets observed in electron microscopy. The small-sized and volatile compounds present in the biochemical extracts are generally eliminated by the procedures of electron microscopy fixation. The heavy molecules (mainly proteinaceous) that are preserved are considered the substratum of the light and potent molecules. As we see it is not possible to follow in cytological terms the different steps of manufacture and extrusion of the secretory products (Noirot and Quennedey, 1974).

A characteristic cytological feature of the class 1 secretory cells is the differentiation of the apical plasma membrane into microvilli that increase the cell surface covered by the secretion. Of course, this increase is dependent on the number and the length of the microvilli; thus, we have found an increasing of the cell surface size by 28 times for *Schedorhinotermes*, 68 times for *Cubitermes*, and 250 times for *Trinervitermes*, whose microvilli are particularly concentrated and long (Quennedey, 1978). Meanwhile, in the frontal gland cells of *Psammotermes* and *Prorhinotermes*, the plasma membrane is simply undulated; this unusual feature was previously only reported in the wax-gland cells of the honeybee.

STRUCTURAL MODIFICATIONS OF THE FRONTAL GLAND CUTICLE

Another noteworthy fact is the existence of structural modifications of the glandular cuticle overlapping the class 1 cells of the frontal glands (Figure 5.16).

With regard to the usual tegumentary cuticle, the most important change concerns the perforation of the cuticle by the epicuticular
pores. Such a minor modification (because at the beginning of the
secretion of the cuticle, the epicuticle is made of small free patches
that become joined after) presents notable functional consequences
that allow the transport of molecules smaller in size than the
epicuticular pore diameter (about 30 nm). Of course, these pores may
be clearly seen on very thin and well-orientated sections. The
glandular cuticle of *Macrotermes* frontal gland is a good example of this
first specialization, the other cuticular structures being unmodified.

A second change is the development of extracellular spaces
between the cell and the cuticle (as in the cephalic part of *Schedorhinotermes* gland) or between two different cuticular layers, endo- or
mesocuticle and epicuticle (as in *Reticulitermes, Cubitermes*, and *Trinervitermes*); or between the inner and the outer epicuticle (as in
Coptotermes). Such spaces permit the storage of the secretory products
regularly secreted by the cells, and various chemical reactions may
occur within them.

This change is often associated with a reduction or the disappearance of some layers. Already in the *Trinervitermes, Cubitermes*, and
Reticulitermes glands, the endo- or mesocuticle is thinner, the pore
canals are absent, and thus the secretion directly diffuses across these
layers. But the reduction may be more marked and the inner layers
could gradually disappear as in the abdominal part of the *Schedorhinotermes* gland, or be lacking altogether. Therefore, in *Coptotermes*, the
cuticle is only composed of separated inner and outer epicuticles
between which the globules of secretion are observed. Such a
simplification phenomenon is at its maximum evolution in the frontal
glands of *Psammotermes, Prorhinotermes*, and *Schedorhinotermes*, in which
the cuticle is only composed of a thin layer, considered to be an outer
epicuticle.

From a functional viewpoint such a slenderizing of the glandular
cuticle allows easier transit of the secretion, but it is difficult to
understand how the slimmer structure plays a protective role for the
cell against the toxic compounds stocked in the glandular lumen. In
the dipteran salivary glands, the glandular cuticle above the secretory
cells has often totally disappeared or has sometimes been reduced to
the outer epicuticular layer (as in the *Macrotermes* salivary acini). Such
a notable modification seems the result of the considerable invagination of the frontal and salivary glands; there is no need to develop a
well-structured cuticle to give mechanical protection.

In conclusion, the capacity for innovation in the defensive
systems developed by termites may prove astonishing. In addition to

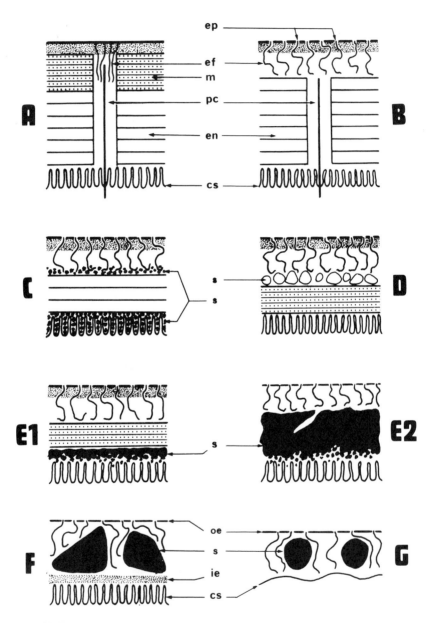

FIGURE 5.16. Structural modifications encountered in the frontal
gland of soldier termites. Compared with usual tegumentary cuticle (A), the
first change is the pitting of the epicuticle by the epicuticular pores (ep), this
feature being common to all frontal glands. In addition, the epicuticular

the side arms comprising varieties of mandibular weapons, the soldiers of termites often also use and sometimes substitute chemical arms in the form of different glandular weapons. These systems are in some cases so extraordinary that they deserve to appear on Hieronymous Bosch, Breughel the Elder, and Jacques Callot paintings (Maeterlinck, 1927). In addition, such ingenuity is also found in the other components of the colony defense system; for example, the defense provided by the fortress-like construction of the nest and also the many integrated defense systems (Deligne *et al.*, 1981).

Acknowledgments

I wish to thank Professor Ch. Noirot for the reading of the manuscript, Dr. Malvyn Eutick and Mrs. Paula Green for the grammar corrections, and my wife, Brigitte Quennedey, for the histological sections and the typing of the manuscript.

REFERENCES

Ampion, M., and A. Quennedey. (1980). The abdominal epidermal glands of termites and their phylogenetic significance. In *Biosystematics of Social Insects* (J. L. Clément and P. E. Howse, eds.). Symposium IUSSI, Paris.

Bacchus, S. (1979). New exocrine gland on the legs of some Rhinotermitidae (Isoptera). *Int. J. Insect Morphol. Embryol.* 8, 135–142.

Bonneville, P. P. (1936). Recherches sur l'anatomie microscopique des Termites. *Arvernia Biol.* 15, 5–127.

Braekman, J. C., D. Daloze, A. Dupont, J. Pasteels, B. Tursch, J.-P. Declercq, G. Germain, and M. Van Meerssche. (1980). Secotrinervitane: A novel bicyclic diterpene skeleton from a termite soldier. *Tetrahedr. Lett.* 21, 2761–2762.

---

filaments (ef) become more abundant and constitute a well-defined layer. These two changes are observed in *Macrotermes* (B). Two other changes occur at the same time; the reduction in thickness of the glandular layers, endocuticle (en), or mesocuticle (m) and the development of extracellular spaces between them as in *Reticulitermes, Trinervitermes* (c), *Cubitermes* (D), and *Schedorhinotermes* (E_1). Such spaces are filled with secretion (s). This reduction may be more marked and the inner layer can disappear entirely in some species. The outer (oe) and inner epicuticles (ie) only are present in *Coptotermes* (F) and the cuticle is composed only of a thin layer considered to be an outer epicuticle in *Psammotermes, Prorhinotermes* (G), and *Schedorhinotermes* (E_2); pc, pore canal with its filament; cs, cell surface of class 1 cells.

Brossut, R. (1973). Evolution du système glandulaire exocrine céphalique des Blattaria et des Isoptera. *Int. J. Insect Morphol. Embryol.* 2, 35–54.

Bugnion, E., and N. Popoff. (1910). Le termite à latex de Ceylan *Coptotermes travians* Haviland; avec un appendice comprenant la description des *Coptotermes destroi* Wasm. et *flavus* nov. sp. *Mem. Soc. Zool. Fr.* 23, 107–123.

Coaton, W. G. H. (1971). Five new termite genera from South West Africa (Isoptera:Termitidae). *Cimbebasia* (A) 2, 1–34.

Cmelik, S. H. W. (1971). Composition of the lipids from the frontal gland of the major soldiers of *Macrotermes goliath.* *Insect Biochem.* 1, 439–445.

Deligne, J. (1965a). Morphologie et fonctionnement des mandibules chez les soldats de Termites. *Biol. Gabonica* 1, 179–186.

Deligne, J. (1965b). Observations sur l'appareil mandibulaire des soldats de Termites: Fonctionnement et histogenèse. Proc. V Congr. IUSSI, Toulouse, pp. 131–142.

Deligne, J. (1970). Recherches sur la transformation des jeunes en soldats dans la société de Termites (Insectes, Isoptères. Thèse Sciences, Université de Bruxelles, Belgique, 424 pp.

Deligne, J. (1971). Mécanique du comportement de combat chez les soldats de Termites (Insectes Isoptères). *Forma et Functio* 4, 176–187.

Deligne, J. (1973). Observations au microscope électronique à balayage d'un nouveau système glandulaire céphalique chez les Termites. Proc. VII Int. Congr. IUSSI, London, 1973, pp. 85–87.

Deligne, J. (1981). Description et affinités de *Verrucositermes hirtus* sp.n. Fonction glandulaire des tubercules du soldat (Isoptères, Nasutitermitinae). *Rev. Zool. Bot. Afr.* (in press).

Deligne, J., and A. Quennedey. (1977). Morphological approach to defensive mechanisms in Termites. Proc. VIII Int. Congr. IUSSI, Wageningen, p. 299.

Deligne, J., A. Quennedey, and M. S. Blum. (1981). The enemies and defense mechanisms of Termites. In *Social Insects* (H. R. Hermann, ed.), Vol. II. Academic Press, New York.

Emerson, A. E. (1945). The neotropical genus *Syntermes* (Isoptera, Termitidae). *Bull. Am. Mus. Nat. Hist.* 83, 430–471.

Emerson, A. E. (1952a). The african genus *Apicotermes* (Isoptera:Termitidae). *Ann. Mus. Roy. Congo Belge* (*Tervuren*) 17, 100–121.

Emerson, A. E. (1952b). The neotropical genera *Procornitermes* and *Cornitermes* (Isoptera, Termitidae). *Bull. Am. Mus. Nat. Hist.* 99, 478–539.

Emerson, A. E., and F. A. Banks. (1965). The neotropical genus *Labiotermes* Holmgren: Its phylogeny, distribution and ecology (Isoptera, Termitidae, Nasutitermitinae). *Am. Mus. Novitates* 2208, 1–33.

Evans, D. A., R. Baker, P. H. Briner, and P. G. McDowell. (1977). Defensive secretions of some African termites. Proc. VIII Int. Congr. IUSSI, Wageningen, pp. 46–47.

Fain-Maurel, M. A., and P. Cassier. (1972). Un nouveau type de jonctions: Les jonctions scalariformes. Etude ultrastructurale et cytochimique. *J. Ultrastruct. Res.* 39, 222–238.

Grassé, P.-P. (1939). Comportement et particularités physiologiques des soldats de termites. *Bull. Soc. Zool. Fr.* 64, 251–262.

Holmgren, N. (1909). Termitenstudien. I. Anatomische Untersuchungen. *Kgl. Svenska Vetenskapsakad Handl.* 44, 1–215.

Howse, P. E. (1975). Chemical defenses of ants, termites and other insects: Some outstanding questions. In *Pheromones and Defensive Secretions in Social Insects* (C. Noirot, P. E. Howe, and G. Le Masne, eds.), Symp. IUSSI, Dijon, pp. 23–40.

Maeterlinck, M. (1927). *The Life of the White Ant.* (English tr. by A. Sutro.) George Allen and Unwin, Ltd., London.

Maschwitz, U., R. Jander, and D. Burkhardt. (1972). Wehrsubstanzen und Wehrverhelten der Termite *Macrotermes carbonarius. J. Insect. Physiol.* 18, 1715–1720.

Maschwitz, U., and Y. P. Tho. (1974). Chinone als Wehrsubstanzen bei einigen orientalischen Macrotermitinen. *Insectes Sociaux* 21, 231–234.

Meinwald, J., G. D. Prestwich, K. Nakanishi, and I. Kubo. (1978). Chemical ecology: Studies from East Africa. *Science* 199, 1167–1173.

Moore, B. P. (1964). Volatile terpenes from *Nasutitermes* soldiers (Isoptera, Termitidae). *J. Insect Physiol.* 10, 371–375.

Moore, B. P. (1968). Studies on the chemical composition and function of the cephalic gland secretion in Australian termites. *J. Insect Physiol.* 14, 33–39.

Moore, B. P. (1969). Biochemical studies in termites. In *Biology of Termites* (K. Krishna and F. M. Weesner, eds.), Vol. I, pp. 407–432. Academic Press, New York.

Noirot, Ch. (1969). Glands and secretions. In *Biology of Termites* (K. Krishna and F. M. Weesner, eds.), Vol. I, pp. 89–123. Academic Press, New York.

Noirot, Ch., and A. Quennedey. (1974). Fine structure of insect epidermal glands. *Annu. Rev. Entomol.* 19, 61–80.

Noirot-Timothée, C., Ch. Noirot, D. S. Smith, and M. L. Cayer. (1979). Jonctions et contacts intercellulaires chez les insectes. II. Jonctions scalariformes et complexes formés avec les mitochondries. Etude par coupe fine et cryofracture. *Biol. Cell.* 34, 127–136.

Prestwich, G. D. (1975). Chemical analysis of soldier defensive secretions of several species of East African termites. In *Pheromones and Defensive Secretions in Social Insects* (C. Noirot, P. E. Howse, and G. Le Masne, eds.), Symp. IUSSI, Dijon, pp. 149–152.

Prestwich, G. D. (1979). Chemical defense by termite soldiers. *J. Chem. Ecol.* 5, 459–480.

Prestwich, G. D., B. A. Bierl, E. D. Divilbiss, and M. F. B. Chaudhury. (1977). Frontal glands of termite *Macrotermes subhyalinus*: Morphology, chemical composition and use in defense. *J. Chem. Ecol.* 3, 579–590.

Prestwich, G. D., M. Kaib, W. F. Wood, and J. Meinwald. (1975). 1,13-tetradecadien-3-one and homologs: New natural products isolated from *Schedorhinotermes* soldiers. *Tetrahedron Letters* 52, 4701–4704.

Prestwich, G. D., J. W. Lauher, and M. S. Collins. (1979). Two new tetracyclic diterpenes from the defense secretion of the neotropical termite *Nasuti-termes octopilis*. *Tetrahedron Lett.* 40, 3827–3830.

Prestwich, G. D., D. F. Wiemer, J. Meinwald, and J. Clardy. (1978). Cubitene: An irregular twelve-membered-ring diterpene from a termite soldier. *J. Amer. Chem. Soc.* 100, 2560–2561.

Quennedey, A. (1973). Observations cytologiques et chimiques sur la glande frontale des termites. Proc. VII Int. Congr. IUSSI, London, pp. 324–326.

Quennedey, A. (1975a). La guerre chimique chez les termites. *La Recherche* 54, 274–276.

Quennedey, A. (1975b). The labrum of *Schedorhinotermes* minor soldier (Isoptera, Rhinotermitidae). Morphology, innervation and fine-structure. *Cell Tissue Res.* 160, 81–98.

Quennedey, A. (1975c). Morphology of exocrine glands producing phero-mones and defensive substances in subsocial and social insects. In *Pheromones and Defensive Secretions in Social Insects* (C. Noirot, P. E. Howse, and G. Le Masne, eds.), Symp. IUSSI, Dijon, pp. 1–21.

Quennedey, A. (1978). Les glandes exocrines des termites. Ultrastructure comparée des glandes sternales et frontales. Thèse Sciences, Université de Dijon, France, 254 pp.

Quennedey, A., G. Brulé, J. Rigaud, P. Dubois, and R. Brossut. (1973). La glande frontale des soldats de *Schedorhinotermes putorius* Sjöstedt (Isoptera, Rhinotermitidae): Analyse chimique et fonctionnement. *Insect Biochem.* 3, 67–74.

Quennedey, A., and J. Deligne. (1975). L'arme frontale des soldats de termites. I. Rhinotermitidae. *Insectes Sociaux* 22, 243–267.

Sands, W. A. (1972). The soldierless termites of Africa (Isoptera:Termitidae). *Bull. Br. Mus. Nat. Hist.* (*Entomol.*) (suppl. 18).

Sannasi, A. (1969). Morphology, histology and histochemistry of the frontal gland of soldier termite *Rhinotermes magnificus* Silvestri (Isoptera, Rhino-termitidae). *La Cellule* 67, 369–375.

Thiéry, J.-P. (1967). Mise en évidence des polysaccharides sur coupes fines en microscopie électronique. *J. Microsc.* 6, 987–1018.

Vrkoc, J., and K. Ubik. (1974). I-nitro-trans-1-pentadecene as defensive compound of termites. *Tetrahedron Lett.* 15, 1463–1464.

Wadhams, L. J., R. Baker, and P. E. Howse. (1974). 4,11-epoxy-cis-eudes-mane: A novel oxygenated sesquiterpene in the frontal gland secretion of the termite *Amitermes evuncifer* Silvestri. *Tetrahedron Lett.* 15, 1697–1700.

Weidner, H. (1974). Beiträge zur Kenntnis der Termiten Angolas, hauptsäch-lich auf Grund der Sammlungen und Beobachtungen von A. de Barros Machado. *Publ. Cult. Comp. Diam. Angola* 88, 13–78.

Wood, W. F., W. Truckenbrodt, and J. Meinwald. (1975). Chemistry of the defensive secretion from the African termite *Odontotermes badius*. *Ann. Entomol. Soc. Am.* 68, 359–360.

6

Elaboration and Reduction of the Venom Apparatus in Aculeate Hymenoptera

Henry R. Hermann

INTRODUCTION

Defensive mechanisms in the eusocial Hymenoptera have played a special role in the rise of this group. Indeed, it has been stated on numerous occasions that one of the very important benefits of gregarious behavior in social organisms is improved defense (Gamboa, 1978; Gibo, 1978; Hermann, 1979; Jeanne, 1982; Seeley, 1982). However significant these mechanisms may be to the rise of sociality, the field of defensive biology in insects has been grossly neglected. As examples, Evans (1958) and Malyshev (1968) (also others, as reviewed by Eickwort, 1981) have both elaborated on significant preadaptations for sociality in the Hymenoptera, and each depicted a sequence of major steps leading to a eusocial existence. These reports mainly emphasize increased complexity in nest construction and nesting behavior. Defensive behavior modifications in solitary and social species were not considered in these reports, although use of the sting in host submission was occasionally discussed. The reason for a lack of information on defensive behaviors in these reports is largely because defensive behaviors in the insects discussed are virtually nonexistent and do not show up in elaborate form until eusociality comes about (however, see Steiner, 1975, 1978).

Prior to the rise of social traits, the ovipositor–sting was employed reproductively in egg deposition, offensively in host paralysis, and defensively in self-preservation. It is obvious that solitary hymenopterous insects did not exhibit a diverse array of defensive behaviors prior to the rise of sociality and they still do not (however, see Steiner, 1975). But social species do, and these behaviors are often very distinctly and elaborately demonstrated (Chadab-Crepet and Rettenmeyer, 1982; Hermann and Blum, 1981; Starr, 1981). The multitude of defensive behaviors that currently exist in the social Hymenoptera (Chadab-Crepet and Rettenmeyer, 1982; Hermann and Blum, 1981; Jeanne, 1979, 1982; Starr, 1981; also see other chapters of this volume) have arisen due to pressures exerted only on insects that exhibit group behavior. Social insects must exhibit defense behavior superior to that of solitary species because they represent a more concentrated food source and their nests are generally easier to find than those of solitary species.

A social existence and a complex defensive behavioral repertoire are therefore strongly interdependent. Indeed, it would be difficult to conceive that sociality would have arisen at all had it not been for the development of a number of protective mechanisms against both vertebrate and invertebrate intruders.

Even though these behaviors appear to all be group related, it seems plausible to think that at least some defensive tactics exhibited by hymenopterous insects at the beginnings of social structure (Carpenter and Hermann, 1979) must have been carried over from similar mechanisms already found in solitary species (Starr, 1981). Escape is ubiquitous in both groups. However, the only overt defensive behavior that appears to be commonly expressed in both solitary and social aculeates, other than the abdomen up-wings spread and wing fluttering territorial behaviors described by Steiner (1975), is the act of stinging, a behavior that was developed even before the rise of aculeates (Hermann and Blum, 1981; Matsuda, 1976).

Most female solitary aculeate hymenopterans are quite capable of delivering a painful sting; however, they are generally reluctant to defend unless directly attacked. It is beneficial for such a solitary female hymenoteran to avoid an encounter rather than run the risk of losing her life in an altercation with an intruder. She has nothing to gain in stinging an intruder unless her life is in jeopardy (Alexander, 1974; Hamilton, 1964, 1972; Lin and Michener, 1972; Maynard Smith, 1978; Trivers, 1971; Trivers and Hare, 1976). However, at that instant, the act of stinging is extremely important in determining her fate and the fate of succeeding generations.

This act of individual defense, so obviously important in solitary species, has been greatly elaborated on in some solitary predators of social insects that subject themselves to an uncommonly high degree of attack, for example, the Mutillidae (Hermann, 1968a). In such cases, the predator often must defend itself against numerous attacking adults of the prey species (Hermann, 1968a), and in doing so, increased manipulative capabilities during the act of stinging are especially important and have led to some interesting venom apparatus modifications (Hermann and Chao, 1983).

This defensive behavior, common to both groups, and employed at the individual level by solitary aculeates, has become associated with numerous other defensive behaviors and has taken on a new meaning of importance in social species. Instead of fleeing from a potential predator or intruder or merely fighting to defend its own life, these insects very often demonstrate altruistic behaviors by defending their queen or the nest in which related progeny are housed, sometimes losing their life in the process (Alexander, 1974; Hamilton, 1964, 1972; Lin and Michener, 1972; Maynard Smith, 1978; Trivers, 1971; Trivers and Hare, 1976). When pressured to defend their own life away from the nest, social insects most often react by escaping (Hermann and Blum, 1981).

With an increasing complexity of defensive behaviors and the rise of additional defensive mechanisms (Hermann and Blum, 1981), certain social hymenopterans have lost their ability to sting and have developed other ways to defend themselves and their colonies. These anagenetic modifications toward an increased complexity of defensive behaviors and the subsequent reductional and elaborational changes in venom apparatus morphology are the main subjects of this chapter.

Description of a Generalized Venom Apparatus

Segmental Origin

Venom sclerites in all aculeate Hymenoptera are homologous to segmental sclerites (eighth and ninth abdominal) and their extensions (D'Rozario, 1942; Gupta, 1950; Gustafson, 1950; Hermann and Blum, 1981; Matsuda, 1957, 1976; Michener, 1944; Oeser, 1961; Scudder, 1957a,b, 1959, 1961a,b; Smith, 1969, 1970; Snodgrass, 1931, 1933, 1935; Stumper, 1960; Zander, 1911).

Prior to the formation of the sting, these sclerites functioned in oviposition, which apparently arose *de novo* in insects before the Carboniferous Period. An ovipositor was well developed in the Carboniferous orders Palaeodictyoptera, Megasecoptera, and Diaphanopterodea (Matsuda, 1976), suggesting that it is indeed an archaic structure. Reduction of the ovipositor has occurred in the Dermaptera, Blattaria, Isoptera, Mantodea, some Neuroptea, Mecoptera, and some Diptera. Total loss has occurred in some Thysanoptera (Tubulifera), Mallophaga, Anoplura, Embioptera and Plecoptera. Its retention, often with some degree of elaboration, is seen in most species of Orthoptera, Phasmida, Grylloblattodea, Hymenoptera, most Psocoptera, most Thysanoptera, Homoptera, and Heteroptera.

According to Matsuda (1976), when reduction of the ovipositor occurred in the Pterygota, the styli and sometimes the second valvifers (ninth gonocoxites) were usually the first to become reduced or lost. The anterior valvulae (from the eighth segment) have persistently been retained in reduced ovipositors of Odonata, Mecoptera, and some nematocerous Diptera.

In adults of the aculeate Hymenoptera, the sclerites are reduced to varying degrees from their most primitive state (Thysanura, Orthoptera) so that they are only partially representative of the plates (terga, sterna, gonocoxites) from which they have arisen (Figure 6.1). The most recent references to homologies and sclerite synonyms may be found in Blum and Hermann (1978a), Hermann and Blum (1981), Kugler (1978), Matsuda (1976), and Smith (1969, 1970).

Venom Sclerites

Paired spiracular plates (Figure 6.1) are representative of the two lateral regions (hemitergites) of the eighth abdominal tergum. The eighth tergum in the lower Hymenoptera is a well-exposed dorsal plate bearing the last abdominal spiracles (Matsuda, 1976). In most Aculeata these plates are mesally connected by a thin posterodorsal bar (Hermann and Blum, 1981). In some more derived social aculeates this mesal connection may only be membranous (Kugler, 1978). As part of the venom apparatus, the paired spiracular plates have retained their spatial relationship with each other, but have moved into the seventh abdominal segment, through a process known as internalization (interiorization of Matsuda, 1976), and now play an important role due to their muscular connections with the rest of the venom apparatus (Goyal, 1977) in venom apparatus manipulation (Blum and Hermann, 1978a; Kugler, 1978).

Paired quadrate plates (Figure 6.1) are remnants of the ninth tergum. The ninth tergum is complete in lower hymenopterous species (Matsuda, 1976) but forms hemitergites in higher groups due to a membranization along the mesal line. These plates and the remaining venom apparatus sclerites have moved, through a second internalization, to lie inside the spiracular plates (Figure 6.2). The positional relationship between quadrate plates and the remainder of the venom apparatus has been retained as in the lower Insecta that possess these structures, although they have been quite reduced from their definitive shape, size, and degree of sclerotization.

Paired oblong plates (Figure 6.1) represent the second gono-coxites on the ninth abdominal segment. They also have undergone considerable reduction in size, shape, and sclerotization. They support paired gonostyli (GO) (i.e., gonoplacs of Scudder) on their posterior extensions. This relationship between the ninth sternum and its styli appears to have been retained in all of the sting-bearing and ovipositor-bearing Insecta. Matsuda (1976), however, points out that the gonostyli in Hymenoptera may not represent true styli. The styli in some Hymenoptera pupae are two-segmented, and it has been suggested that only the apical, conical portion represents the original styli.

The anterior region of each oblong plate supports a hair plate (Hermann and Douglas, 1976a,b) with numerous trichoid sensilla that extend from the dorsal surface of the oblong plate toward the posteroventral articulating end of the triangular plate. The gonostyli also have numerous sensilla trichodea extending from much of their lateral surface (Hermann and Douglas, 1976a,b; Kugler, 1978).

Paired fulcral arms (Figure 6.1) represent only small parts of the ninth sternum. These structures are referred to by Matsuda (1976) as the ventral processes; as such, they may represent basal bifurcations of the first valvulae or remnants of the eighth sternum. The base of each of the fulcral arms, which articulates on each side to the sting base, represents the pivot point for sting depression and levation. Sting depression and levation (Figure 6.2), along with rotation or pivoting (Hermann and Chao, 1983), are the main movements behind direct sting manipulation. The mesobasal region of the fulcral arms supports several sensilla placodea (Hermann and Douglas, 1976a,b) that probably function during sting depression, levation, rotation, or pivoting.

Controversy exists over the origin of the triangular plates and furcula (Figure 6.1). According to Scudder (1957a,b, 1959, 1961a,b, 1964), each of the paired triangular plates is a gonangulum (also reviewed by Matsuda, 1976) that has arisen from precoxal plates of the ninth segment. Sharov (1966) prefers to consider the triangular plates as arising as vestiges of the ninth sternum that were later pushed forward by the base of the second pair of valvifers (oblong plates). Speculation as to their origin also has included the eighth sternum (Snodgrass, 1931, 1933, 1935). Movement of the triangular plates functions in alternate lancet extension and retraction during the act of stinging.

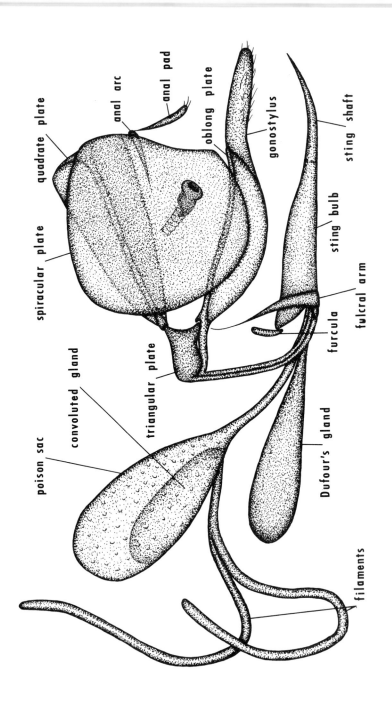

quadrate plate

anal arc

anal pad

oblong plate

gonostylus

sting shaft

spiracular plate

sting bulb

fulcral arm

furcula

convoluted gland

triangular plate

Dufour's gland

poison sac

filaments

FIGURE 6.1. Generalized venom apparatus, showing slightly ascendant sting shaft, as found in primitive ant species of the Myrmeciinae and Ponerinae. The spiracular plates (8th hemitergites) surround the upper portions of the remaining venom apparatus sclerites, due to a second internalization. The venom reservoir (poison sac in this illustration) and its components (convoluted gland and filamentous glands) are also typical of species in these primitive ant subfamilies, with a single filament parting from the proximal end of the sac before branching.

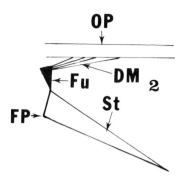

FIGURE 6.2. Diagrammatic lateral view of sting (St), furcula (Fu) and oblong plate (OP), pointing out the fulcral point (FP) during contraction of the sting depressor muscles (DM) which originate on the oblong plate and insert on the distal region of the furcula. (Redrawn from Hermann and Chao, 1983.)

Flemming (1957), Oeser (1961), and Trojan (1935) derived the furcula as an apodeme of sternite 9 while Zander (1899) derived it from sternite 10. Hermann and Chao (1983; also see following discussion) have shown that the furcula has arisen as an elaboration of the anterior tip of the sting bulb, the latter structure being formed from gonapophyses at the bases of the ninth segment.

The primary region of the sting (Sting Shaft, Figure 6.1) is represented by fused gonapophyses from the ninth segment, and the paired lancets (Figure 6.3), representing gonapophyses from the eighth segment. In transverse section (Figure 6.4), the lancets comprise only a small ventromesal portion of the sting shaft. They move alternatingly anteriad and posteriad by way of tongue and groove articulations (rachis and aulax, respectively, of Smith, 1970) on the main body of the sting shaft. The dorsally and laterally positioned, fused second valvulae and the ventrally positioned pair of lancets surround the venom canal through which venom flows from the venom reservoir to the sting victim.

The dorsal (second) valvulae of the sting are connected, by way of the second rami, to the paired oblong plates. These rami support a number of trichoid sensilla on their posterior border (Hermann and Douglas, 1976a,b). The lancets connect, by way of the first rami, to the triangular plates. The first rami support serrations that rub on the trichoid sensilla of the second rami.

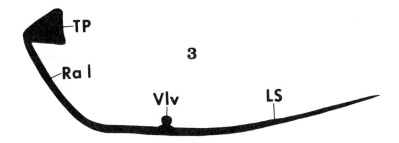

FIGURE 6.3. Lateral view of a lancet as found in most stinging aculeates, with a valve (Vlv) found near the proximal end of the lancet shaft (LS). Ra 1, first ramus; TP, triangular plate.

Origin and Significance of the Furcula

Description

The furcula (Figures 6.5–6.7) as briefly stated in the previous section, is a small sclerite that articulates to the dorsal region of the anterior sting base. The shape of the furcula varies in hymenopterous insects from Y-shaped, U-shaped, and V-shaped (e.g., Vespidae; Pseudomyrmecinae, Myrmicinae, Formicidae) (Figures 6.5 and 6.6) to X-shaped (in the Ponerinae, Myrmeciinae, Formicidae) (Figure 6.7). This furcula–sting base articulation is dicondylic (Figure 6.8), forming a transverse axis of motion for furcular movements. Synonyms for the term furcula, as listed by Hermann and Blum (1981) and Kugler (1978), are as follows: apodeme of stylet, detached notum, fourchette, Gabelbein, intervalvula.

Function

Anteroventral and posterodorsal movements of the furcula effect levation and depression of the distal sting tip, respectively. Two antagonistic sets of muscles influence sting depression and levation. A pair of muscles originating on the mesal side of each oblong plate and inserting on the ventrolateral edge of the furcula (posterior gono-coxapophyseal muscles), functions in sting depression (DM, Figure 6.2). Antagonistic muscles, the sting levators (anterior gonocoxa-pophyseal muscles), originate on the posterior region of the second rami and insert on the anterior tip of the sting base (LM, Figure 6.2).

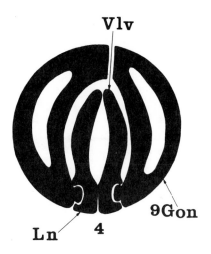

FIGURE 6.4.   Transverse section through sting (as in *Paraponera clavata*, Formicidae, Ponerinae) in region of lancet valves (Vlv), showing fused second valvulae (9th gonapophyses, 9 Gon) and lancets (Ln) that move on one another by tongue and groove articulations (rachis and aulax, respectively, of Smith, 1970).

At the onset of hymenopteran evolution, strong articulation between sclerites of the ovipositor system did not allow diverse manipulative capabilities. As is discussed shortly, however, reduced sclerotization between the various plates led to increased maneuverability, and the posterior gonocoxapophyseal muscles increased their functional significance to pivot or rotate the ovipositor and sting.

Evolution of Furcula from a Nonfurculate, Nonstinging Hymenopteran

A furcula is lacking in sawfiles (Figure 6.9) (Smith, 1970); however, ovipositor levators (OL) and depressors (DM) are present. Depressor muscles insert on the ovipositor base rather than on a furcula. Thus the furcula has not yet been developed in insects representing this stage of hymenopteran evolution. This condition in the Symphyta, in which the ovipositor base has one pair of muscles inserting on it from their points of origin on the oblong plates (ninth gonocoxites), represents the definitive state in the Hymenoptera

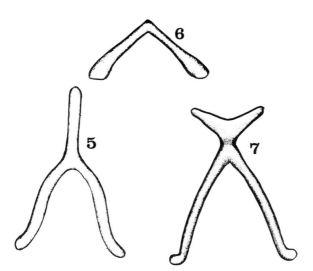

FIGURES 6.5–6.7. Furculae in species of ants. (5) *Pseudomyrmex*; (6) *Pogonomyrmex*; (7) *Myrmecia*. (Redrawn from Hermann and Chao, 1983.)

from which all derived conditions have arisen (Hermann and Chao, 1983).

This condition prior to the development of a furcula apparently has persisted phyletically through all of the ovipositor-bearing Insecta other than the Hymenoptera (Hermann and Chao, 1983; Matsuda, 1957, 1958) and most hymenopterous parasitoids, apparently changing just prior to or during the rise of aculeates (Hermann and Chao, 1983; Hermann and Morrison, 1980). Muscles homologous and analogous to those on the ovipositor base in the Symphyta have been reported in some parasitoids (Hanna, 1934) while a furcula has been reported in others (Hermann and Morrison, 1980).

Throughout the Aculeata, a furcula persists, except in certain formicids and other derived hymenopterous groups that have become specialized in other types of defense (Hermann and Chao, 1983; Kugler, 1980). In these insects, the furcula has been reduced, but predecessors of these insects were furculate. In spite of the changes that have evolved along with furcular elaboration, the definitive muscle groups associated with ovipositor and sting levation and depression have retained their points of origin and insertion but have changed their significance in the maneuverability of these gono-coxapophyseal extensions.

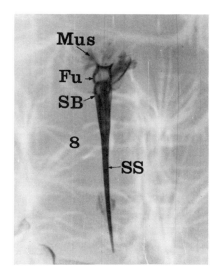

FIGURE 6.8. Sting and furcula of *Odontomachus* sp. Furcula (Fu) is basically X-shaped. Mus, muscle inserting on distal region of furcula from its origin on oblong plate; SB, sting base; SS, sting shaft.

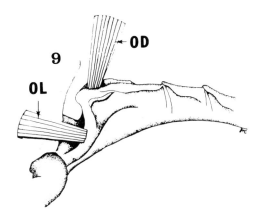

FIGURE 6.9. Ovipositor base in a sawfly, showing the ovipositor depressor (OD) and its antagonistic muscle, the ovipositor levator (OL). These muscles are better known as the posterior and anterior gonocoxapophyseal muscles. (Redrawn from Smith, 1970.)

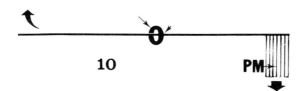

FIGURE 6.10. Diagrammatic dorsal view of sting, showing pivoting muscles (PM) and their effect on a bilateral movement of the sting tip (arrows).

Some degree of elaboration occurred in the ovipositor or sting depressor muscles prior to the time that the furcula was arising. These muscles lined the mesal side of the oblong plate so that when reduction of the venom apparatus sclerites came about, resulting in a decrease in sclerotization and increased movement between these sclerites, the sting depressors took on an additional function, that of pivoting the sting base from left to right (Figure 6.10). This increased manipulative capability probably occurred before furcular development in the aculeate predecessor, paving the road for the use of this apparatus in a defensive capacity.

Further change resulted in the development of a fold between the anterior sting base extension, to which the muscles of pivoting are attached, and the sting bulb, resulting in the production of the furcula (Figure 6.11). Instead of being positioned in the same plane with the sting, the furcula stands upright from its dicondylic attachment with the sting bulb (Figure 6.2). Through part of the muscles from the oblong plates, the more posterior ones, sting depression remains much the same as it did prior to furcular elaboration. However, the anteriormost portion of these muscles, moving from their point of origin on each oblong plate and moving directly mesad to insert on the laterodistal region of the dorsal furcular extension, effects a sting rotation rather than a pivoting motion (Figure 6.11). This represents an additional step toward an increase in manipulative capability in stinging Hymenoptera. Through depression, levation, and sting rotation, this defensive weapon shows almost unlimited maneuverability. This is the condition common to those species that were evolving to be aculeates.

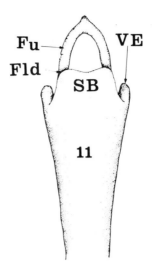

FIGURE 6.11. Sting base (SB) showing the anterior sting base extension in the form of a furcula (Fu) and the point of folding (Fld) that enabled the furcula to move to a position upright on the sting base during the rise of a furcula. VE, ventral sting base extension.

ELABORATION

Introduction

Since the rise of a furcula (Hermann and Chao, 1983) in an aculeate predecessor, there have been no new sclerites added to this system. The following are some of the known elaborational changes that have come about:

Gonostyli (also Gonoplacs)

The gonostylar lobes in ponerine and myrmeciine (Hermann and Blum, 1981; Hermann *et al.*, in press 1984) ant species (not including *Nothomyrmecia macrops*, Kugler, 1980) are bilobed (PL and DL, Figure 6.12), the proximal lobe being the larger of the two. The distal lobe supports a number of sensory hairs (Hermann and Douglas, 1976a,b; Kugler, 1978, 1980). Gonostyli in mutillids are also bilobed (Figures 6.13 and 6.14), but the distal portion is largest and the

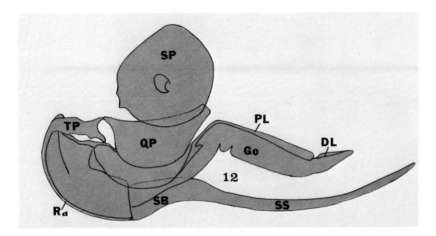

FIGURE 6.12. Lateral view of the venom apparatus of *Dinoponera grandis*, showing the bilobed condition of the gonostyli. DL, distal gonostylar lobe; Go, gonostyli; PL, proximal gonostylar lobe; QP, quadrate plate; Ra, rami; SB, sting base; SP, spiracular plate; SS, sting shaft; TP, triangular plate. (Redrawn from Hermann *et al.*, in press 1984.)

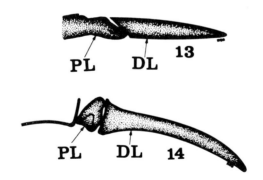

FIGURES 6.13 and 6.14. Mutillid gonostyli. (13) *Myzinum obscurum*; (14) *Timulla ornatipennis*. DL, distal lobe. PL, proximal lobe.

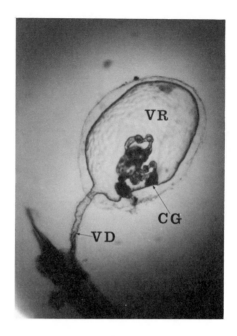

FIGURE 6.15. Venom reservoir (VR) in *Atta texana*. CG, convoluted gland. VD, venom duct, leading to sting base (Courtesy of John Moser).

proximal lobe supports most of the trichoid sensilla (Hermann, 1968a). Kugler (1978) indicated that there is an apparent bilobity in the gonostyli of some myrmicines (e.g., *Myrmica*, Figure 11, in Kugler, 1978). However, the gonostyli of most myrmicine species is uni-lobular.

Venom Apparatus Glands

There has been an increased complexity of the glandular components of the venom apparatus. The convoluted gland associated with the venom gland in most aculeates has sunken into the venom reservoir (Figure 6.15). In its most primitive form (e.g., in bees, Figure 6.16), the convoluted gland extends from the apical end of the venom reservoir as a single filament and sometimes branches distally (Owen and Bridges, 1976). In most aculeate wasps, the convoluted gland is inside the venom reservoir and exits from the venom reservoir's apex to branch immediately into two or more filamentous glands. The convoluted gland in all ants, except the Formicinae, is

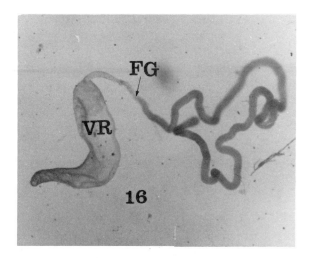

FIGURE 6.16. Venom producing and storing components in *Apis mellifera*. FG, filamentous gland extending from reservoir apex; VR, venom reservoir.

inside the venom reservoir. In ponerine, myrmeciine, and pseudo-myrmecine ants, a single filament extends from the sac's basal region and branches into two filaments some distance from the reservoir. In myrmicine ants, paired filamentous glands extend from near the reservoir apex, some of which are somewhat globate (Kugler, 1978). These glands in the Dolichoderinae extend from near the reservoir's apex in paired form, and each is distinctly globate in general appearance rather than filamentous. The convoluted gland in formicine ants is typically outside the venom reservoir.

Dufour's gland varies only slightly in general appearance in all of the Hymenoptera. Its most primitive and most common form in aculeates is a simple saclike structure (Figure 6.16). In some formicines and halictines (Hermann and Blum, 1981; Wilson and Regnier, 1971), Dufour's gland has become globate or bilobed (Figure 6.17).

The venom reservoir is often surrounded by a relatively sparse musculature in aculeates. However, the reservoir in all of the wasps of the superfamily Vespoidea appears to have an extensive muscle supply (Figure 6.18) (Hunt and Hermann, 1970), an elaboration that apparently occurred along with an important reductional change in the venom sclerites. Lancet valves, typically on the dorsal side of the anterior region of the lancet shaft in all other stinging aculeates (Figure 6.3), are lacking in the Vespidae (Figure 6.19). Lancet valves

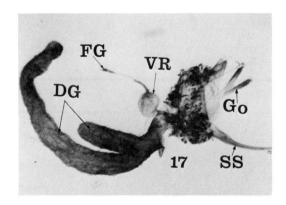

FIGURE 6.17. Venom apparatus of *Halictus ligatus*, showing enlarged Dufour's gland (DG). FG, filamentous gland; Go, gonostyli; SS, sting shaft; VR, venom reservoir. (Photo courtesy of A. Hefetz.)

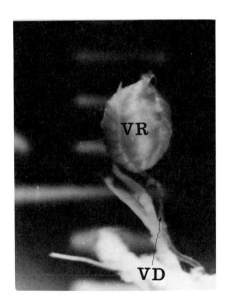

FIGURE 6.18. Venom reservoir (VR) in *Polistes annularis*, showing extensive muscle supply around reservoir, which is characteristic of the vespids. VD, venom duct leading to sting base.

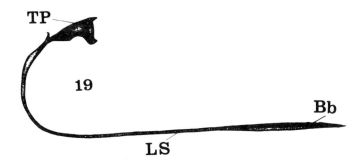

FIGURE 6.19. Lateral view of vespid lancet, showing lack of valve (compare with Figure 6.3). Bb, lancet barbs; LS, lancet shaft; TP, triangular plate.

are routinely considered to function in facilitating venom evacuation through the venom canal in the sting of most aculeates. However, in the Vespidae it is a lack of valves that facilitates a rapid venom evacuation. Contraction of the powerful muscles surrounding the venom reservoir forces venom through the sting shaft at a rapid rate. Lancet valves would only slow such a forceful flow. Venom flow in this group, therefore, does not depend on alternating lancet movements (Hunt and Hermann, 1970).

Furcula

The furcula has taken on many forms in aculeates (Hermann and Chao, 1983; Kugler, 1978, 1980) (Figures 6.5–6.7) and in certain solitary and social species in which stinging behavior is an important defensive mechanism, the dorsal furcular apodeme has become longer (DA, Figure 6.8) and functions in a more efficient manner in sting rotation. Such an increased efficiency has reached its peak in the Mutillidae (to be discussed next) (Hermann, 1968a).

Lancet Barbs

Increased size of the lancet barbs (Figure 6.20) represents a distinctly independent phyletic development in three different aculeate groups: *Pogonomyrmex* ants, bees of the genus *Apis*, and certain tropical wasps. As such, increased barb size represents a clear-cut case

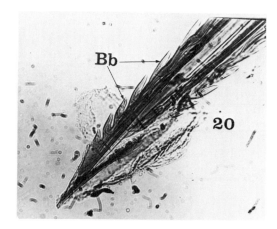

FIGURE 6.20. Tip of lancets in *Apis mellifera*, showing enlarged, sting autotomous barbs (Bb).

of convergent evolution. The associated behavior, sting autotomy, represents an extremely well-defined case of altruism (Hermann and Blum, 1981) (see Sting Autotomy).

Fulcral Arms

In a limited way there may have been some increase in the size of the fulcral arms (Figure 6.21) in mutillid wasps. Since this group has had distinct elaborational changes in the sting shaft, it is believed that the fulcral arms in this group may have developed in relationship to the function of sting depression.

Sting Shaft

As mentioned above, sting shaft elaboration has come about in the form of descendance or ascendance. Ascendance (see sting ascendance in glossary for explanation) of the sting is common among the primitive ants (Hermann, 1969; Kugler, 1980; Robertson, 1968). Strong ascendance is seen in the sting of *Onychomyrmex hedleyi* (Ponerinae) (Hermann, 1969) (Figure 6.22). Sting descendance has been found in all vespids (Hunt and Hermann, 1970; Hermann and Krispyn, 1975; Spradbery, 1973) and many solitary wasp species (Hazeltine, 1967) (Figures 6.23 and 6.24). Kugler (1978) reports some sting descendance in myrmicine ants with a Grade 2 venom apparatus. Descendance reaches its peak in the coiling sting of mutillid wasps (Figure 6.21).

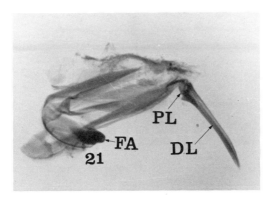

FIGURE 6.21.   Lateral view of venom apparatus of *Dasymutilla v. vesta* (specimen courtesy of H. O. Evans), showing enlarged and darkly sclerotized fulcral arm (FA). DL, distal gonostylar lobe; PL, proximal gonostylar lobe.

Increased Sclerotization

Increased sclerotization has taken place in some vespid species, especially in the genus *Polistes* (Hunt and Hermann, 1970) (Figure 6.24). However, desclerotization is more often the general rule, especially in the ants.

Sting Coiling in the Mutillidae

The venom apparatus in certain solitary Hymenoptera is especially well developed. A prime example is found in the Mutillidae, reported by Hermann (1968a) in *Dasymutilla occidentalis* (L.) and well demonstrated in other species as well (Figure 6.25).

Description

In *Dasymutilla occidentalis*, the sting is distinctly exsertile but sometimes not evident because it is covered laterally by the gonostylar processes (gonoplacs). The basal region of the venom apparatus, including the triangular, oblong, quadrate, and spiracular plates (Figure 6.25), is similar in general arrangement to most other stinging aculeates. However, the remaining sclerites have been modified considerably.

Looping of Sting

Rami (Ra) from the paired triangular plates (TP, Figure 6.25) pass ventrad as in other hymenopterans but then loop strongly anteriad

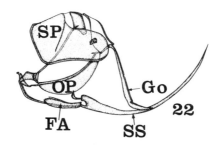

FIGURE 6.22. Lateral view of venom apparatus of *Onychomyrmex hedleyi*, showing extreme sting ascendance. FA, fulcral arm; Go, gonostylus; OP, oblong plate; SP, spiracular plate; SS, sting shaft.

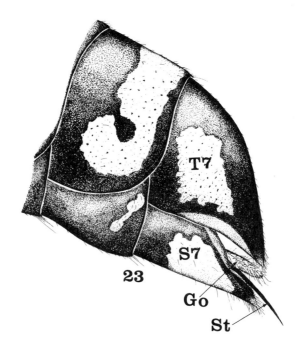

FIGURE 6.23. Lateral view of abdominal tip of *Vespula maculata*, showing exsertile, descendant sting. Go, gonostylus; St, sting; S7, seventh sternite; T7, seventh tergite. (Redrawn from Hermann and Krispyn, 1975.)

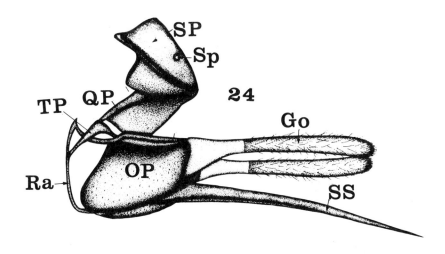

FIGURE 6.24. Lateral view of venom apparatus of *Polistes annularis*, showing heavy sclerotization of most venom apparatus sclerites and descendant sting. Go, gonostylus; OP, oblong plate; QP, quadrate plate; Ra, rami; Sp, spiracle; SP, spiracular plate; SS, sting shaft; TP, triangular plate. (Redrawn from Hunt and Hermann, 1970.)

behind the fulcral arms (FA) before entering the sting. Rami from the oblong plates do the same. The sting itself has been looped dorsad and anteriad so that the sting base (SB) extends anteriad for a short distance, bending dorsad, then posteriad, and finally posteroventrad so that the distal sting shaft (SS) lies between the paired gonostylar extensions (GO). In short, the sting assumes a highly descendant, coiled shape. The lancets that ride on the mesal tonguelike processes of the sting shaft dorsum follow the same curvature. Such an abrupt looping of these structures makes it difficult to move the lancets inside the sting unless the sting is at least slightly uncoiled first.

Gonostylar Functions

The gonostyli are fused with the oblong plates and with each other at their bases (Fus). They are well sclerotized and function in the Mutillidae in three distinct ways: (1) as sensory devices during host findings (Lamborn, 1915); (2) as the two lateral halves of a sting

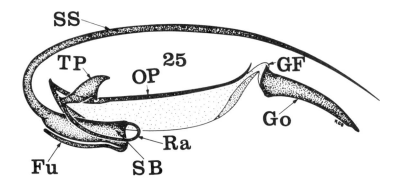

FIGURE 6.25. Lateral view of venom apparatus of *Pseudomethoca ocealla*, showing coiled sting and elongate furcula (Fu). Details of gonostylar bilobity have been omitted. GF, gonostylar fulcrum, important in sting manipulation; Go, gonostylus; OP, oblong plate; Ra, coiled rami; SB, sting base; SS, sting shaft; TP, triangular plate.

sheath; and (3) as a guide for sting maneuverability (further explanation below).

Furcular Elongation

The furcula is extremely long and slender (Figure 6.32), following the sting base through much of the basal portion of its coil (Figure 6.17). Because of the presence of a long dorsal furcular apodeme (DA, Figure 6.18), the muscles of depression and rotation are more efficient in sting maneuverability (Hermann and Chao, 1983). Simultaneous depression and rotation uncoil the lengthy sting and increase both directional maneuverability and length of thrust.

Venom Apparatus Maneuverablility

During contraction of the posterior gonocoxapophyseal muscles, effecting sting depression and rotation (RM, Figure 6.26), the channel provided at the gonostylar base (GB) in mutillids increases sting maneuverability by functioning as a floating fulcral point (PP) for lateral sting movements (Figure 6.26). At rest, the force arm (FA) of the sting lever system (basal portion between sting base and gonostylar fulcrum) (GF, Figure 6.25; GB, Figure 6.26) is very long as compared to the resistance arm (RA) (from gonostylar fulcrum to

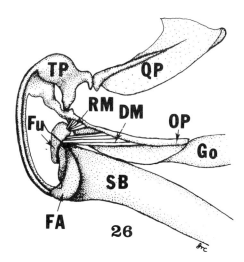

FIGURE 6.26. Lateral view of sting base of *Paraponera clavata*, showing rotating (RM) and depressor (DM) muscles inserting on furcula (Fu) from their point of origin on the oblong plate (OP). FA, fulcral arm; Go, gonostylus; OP, oblong plate; QP, quadrate plate; SB, sting base; TP, triangular plate.

distal sting tip). However, full contraction of the depressor muscles uncoils the sting posteriad, increasing the length of the resistance arm and decreasing the length of the force arm. Limited contraction of the depressor muscle results in varying degrees of sting extension through the gonostylar fulcrum, changing the mechanical advantage of the lever system (Hermann and Chao, 1983). Maneuverability of thrust direction was noted early by Schoenfeld (1878) when he stated that on being extruded, the sting rapidly bends dorsad, ventrad, and laterad and is capable of moving throughout a circular path.

Such maneuverability is important to mutillids in their predatory habits. They often enter bumblebee or honeybee colonies and deposit eggs on the immatures. In doing so they often encounter strong hostility from defending adult bees, sometimes killing hundreds of them.

Overview

In light of the coiled sting, elongated furcula and constantly changing fulcral point for sting maneuverability, the venom apparatus in mutillid wasps is extremely well developed as a defensive mechanism. In structural and functional elaborations, it represents an extreme in venom apparatus development, the most effective defensive system known in hymenopterous insects.

Sting Autotomy (= Autothysis)

Sting autotomy, found only associated with certain social hymenopterous insects, has been witnessed in a number of vespids, all members of the genus *Apis*, and the ant genus *Pogonomyrmex* (Hermann, 1971; Hermann and Blum, 1981; Maschwitz, 1964; Maschwitz and Kloft, 1971). Sting autotomy results from lancet barbs that have become so large that the insect that has such a condition has difficulty in removing them from the sting victim (Figure 6.20). This results in a severing of the distal gastral tip from the body between the sixth and seventh abdominal segments. The sting and its muscles and glands remain with the severed region, continuing to function in venom injection. An insect that sting autotomizes dies after losing its sting.

A number of species of wasps listed by Hermann and Blum (1981) do not sting autotomize in human skin. However, sting autotomy also depends on a factor other than those related to barb size, that is, density of the material being stung. In addition to the sting autotomous wasps listed by Hermann and Blum (1981), W. L. Overal, in Belem, Brazil, has found *Polybia rejecta, Brachygastra lecheguana*, and *Protopolybia pumila* to sting autotomize (pers. comm.).

In addition to the lancet barbs, Weiss (1978) reported well-developed barbs on the distal sting tip (stilet barbs) of some apids (four in *Apis florea*, three to four in *A. dorsata* and *A. cerana*). *Apis mellifera* does not have well-developed barbs on the sting tip. Eleven barbs were found on each lancet in *A. dorsata, A. mellifera*, and *A. cerana*. Ten were found in *A. florea*. Lancet barbs were best developed in *A. florea*, followed by *A. dorsata* and *A. mellifera*, and poorly developed in *A. cerana*.

Increase in Defensive Behaviors

As stated by Kugler (1978), the ramifications of the loss of stinging ability in the biology and morphology of many ant species are far reaching. He listed several compensatory adaptations for defense in nonstinging ants, for example, modifications of the mandibles, chemical defense, and attack behaviors. To these adaptations we should add many of the anatomical and behavioral modifications listed by Hermann and Blum (1981).

REDUCTION

Introduction

Reductional changes in the venom apparatus of aculeate Hymenoptera are more obvious than elaborational ones. Brief mention of some reductional changes (e.g., loss of lancet valves in the Vespidae) has already been made. Several other changes that have come about are listed in the following sections.

Desclerotization

Desclerotization has commonly occurred in the Formicidae and other hymenopterous groups as well. Desclerotization and reduction of sting sclerites are closely related. Both have taken place primarily at the sting tip and gonostyli. Sting disfunction and membranization of the sting tip have occurred in some Myrmicinae (Hermann and Blum, 1981; Kugler, 1978) (Figure 6.27).

Kugler outlined four basic grades of venom apparatuses in the Myrmicinae, based on general appearance and degree of sclerotization. Grade 1 has most of the characteristics of a hypothetical ancestor, chosen in that paper to be *Amblyopone pallipes*. The venom apparatus is well sclerotized and sclerites are large with well-developed muscle insertions. The sting shaft is ascendant, the valve chamber, sting bulb, and sting base are relatively low, and the sting is sclerotized well enough to pierce arthropod cuticle.

A Grade 2 venom apparatus is also used in stinging, but it is not as heavily constructed; the widths of the quadrate, triangular, and spiracular plates and posterior arm of each oblong plate tend to be reduced along with reduced muscle insertions. The gonostyli are decreased in length, sclerotization, and setation, and its proximal and distal segments tend to fuse. Barbs are lost on the lancets. The sting is straight or descendant.

A Grade 3 venom apparatus has a more enlarged valve chamber and sting bulb. The sting base and furcula are strongly arched. The dorsal furcular arm and the basal ridge of the sting base are reduced or lost while the basal notch is longer. The sting is poorly sclerotized and

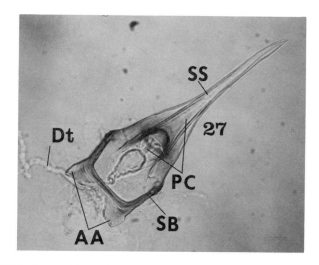

FIGURE 6.27. Ventral view of sting of *Atta texana*, showing membranous sting tip and anterolateral sting base apodemes (AA) that attach to ventral furcula arms. Dt, venom duct from reservoir to sting base; PC, venom canal; SB, sting base; SS, sting shaft.

has lost its piercing function. The distal tip of the lancets and sting shaft may become spatulate.

A Grade 4 venom apparatus is weakly sclerotized. The arms of the oblong plates tend to fuse. The spiracular and quadrate plates intermesh rather than broadly overlap. The gonostyli are shorter, wider, more weakly sclerotized and with fewer sensilla. The triangular plate may be larger than in more primitive species. The valve chamber is highly reduced or absent. The sting bulb is less arched. The furcula is roughly bar shaped to partially or completely membranous.

Sting disfunction and membranization are evident in the Dolichoderinae and developed to an extreme in the Formicinae. Gonostylar reduction is most pronounced in these same groups. Along with sting reduction, reduction in lancet barbs and lancet valves is apparent in the latter groups listed above. Kugler put the Dolichoderinae into Grade 4, mentioning that the Formicinae would represent an even more advanced grade of evolution.

Sensory Structures

With desclerotization and general venom apparatus reduction, the number of sensory structures has been reduced (Hermann and Douglas, 1976b; Kugler, 1978).

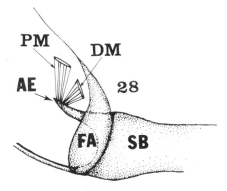

FIGURE 6.28. Lateral view of sting base of *Simopelta oculata*, showing pivoting (PM) and depressor (DM) muscles of sting base. AE, anterior sting base extension; FA, fulcral arm; SB, sting base. (Redrawn from Hermann, 1968c.)

The furcula has undergone a number of changes, including some size reduction and possible loss in the following groups of ants.

Fusion between Sting Base and Furcula

The furcula has fused to the sting base in *Simopelta oculata* (Ponerinae) (Hermann, 1968c) (Figure 6.28) and all of the Cerapachiinae (Figure 6.29), Dorylinae (Hermann, 1969) (Figure 6.30), Aneuretinae, Dolichoderinae (Hermann, 1968b; Blum and Hermann, 1978b) (Figure 6.31), and possibly in some of the Myrmicinae (Kugler, 1978). The remainder of the Ponerinae (Hermann, 1969), all Myrmeciinae, Pseudomyrmecinae, and most Myrmicinae are to date known to be furculate. Prior to Brown's treatment of the cerapachyines (1975) in which he includes them as three tribes in the subfamily Ponerinae, *Simopelta oculata* was the only ponerine in which the furcula had refused to the sting base (Hermann, 1968c). However, *Cerapachys* and *Acanthostichus* are known to have this same postfurculate condition (Hermann, 1969). Why this and others of the former group have lost their furcula is partially an unanswered question. However, its loss appears to accompany some decrease in sting efficiency and modification of their defensive arsenal.

The sting base in *Eciton, Neivamyrmex, Cheliomyrmex* and *Nomamyrmex* (Figure 6.30) all have a well-defined anterior extension (Hermann, 1969). The anterior end of the sting base in *Labidus* appears to be somewhat truncate. Species of *Dorylus* all have broad

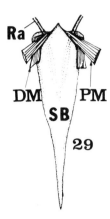

FIGURE 6.29. Dorsal view of *Syscia typhla*, showing pivoting (PM) and depressor (DM) muscles of sting base. Ra, ramus; SB, sting base. (Redrawn from Hermann, 1969.)

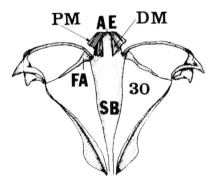

FIGURE 6.30. Dorsal view of venom apparatus of *Cheliomyrmex morosus*, showing pivoting (PM) and depressor (DM) muscles of sting base (SB). AE, anterior sting base extension; FA, fulcral arm. (Redrawn from Hermann, 1969.)

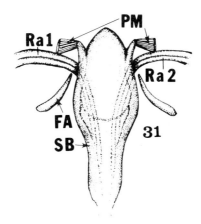

FIGURE 6.31.   Dorsal view of sting base of *Aneuretus simoni*, showing pivoting muscles (PM) inserting on sting base (SB). FA, fulcral arms; Ra 1, first ramus; Ra 2, second ramus. (Redrawn from Hermann, 1968b.)

sting bases. Subgenera of the genus *Dorylus* all possess anterolateral extensions instead of an extension that is more mesally enlarged.

Muscles that rotate the sting in furculate species attach to the sting base (Figure 6.32) and pivot it in these dorylines (Hermann and Chao, 1983). Depressor muscles also insert on the sting base.

When the venom duct enters the sting bulb dorsad of Dufour's gland, it adheres to the bulb's dorsal wall. The venom duct is opened by a dilator muscle inserting on its ventral side, and it is closed by a zygomatic occlusor muscle that originates on the mesal wall of the sting base (Figure 6.33).

Loss of Sting Function (*Aneuretus* and Dolichoderines)

Aneuretus simoni, the only living species in the ant subfamily Aneuretinae (most recently included in the Dolichoderinae—Snelling, 1981), was at one time thought to be linked to both the Ponerinae and Dolichoderinae (Hermann, 1968b). One of the important incriminating features linking this species to the ponerines was the exsertile sting.

Most of the sting in *Aneuretus* is relatively well sclerotized when considering the size of the workers and comparing them to dolichoderines. The proximal portion forms a relatively large bulb, the

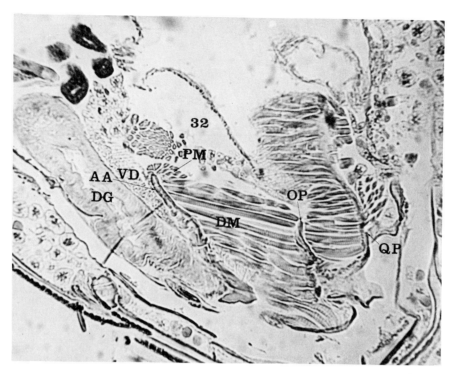

FIGURE 6.32. Sagittal section through sting base of *Eciton hamatum*, showing depressor (DM) and pivoting (PM) muscles inserting on anterior sting base extension (AA) from their point of origin on oblong plate (OP). DG, Dufour's gland; QP, quadrate plate; VD, venom duct.

anterior extension of which is somewhat membranous. There is no furcula. Fusion has occurred between it and the sting base.

The paired gonostyli are unilobular in *Aneuretus*; this character separates this species from the Ponerinae who have long, well-defined, bilobed gonostyli. In general appearance of the gonostyli and sting base, *Aneuretus* is distinctly dolichoderine-like and represents a state of reduction in which the venom apparatus is losing or has lost its importance as a defensive device. The dolichoderine venom apparatus (Figure 6.34) represents a more advanced state of reduction in which the apparatus is less sclerotized and nonfunctional as a sting.

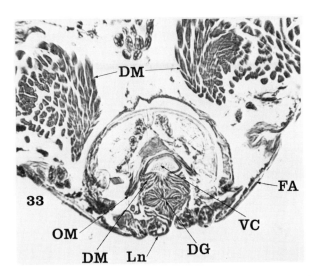

FIGURE 6.33. Transverse section through sting base of *Paraponera clavata* (Formicidae, Ponerinae). DG, Dufour's gland; DM, dilator muscle of venom canal; FA, fulcral arm; Ln, lancet; OM, zygomatic muscles of venom canal from mesal surface of fulcral arms; VC, venom canal.

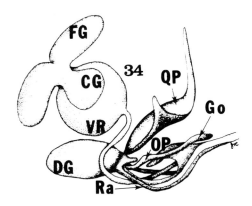

FIGURE 6.34. Lateral view of venom apparatus of *Iridomyrmex pruinosus*, showing reduced sclerites and globate filamentous glands (FG). CG, convoluted gland; DG, Dufour's gland; Go, gonostyli; OP, oblong plate; QP, quadrate plate; Ra, ramus; VR, venom reservoir. (Redrawn from Hermann and Chao, 1983.)

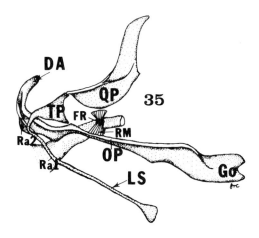

FIGURE 6.35. Lateral view of venom apparatus of *Camponotus pennsylvanicus*, showing reduced first valvulae (lancet shaft, LS), loss of second valvulae, and retention of a furcular remnant (FR). DA, dorsal apodeme of oblong plate; Go, gonostylus; LS, lancet shaft; OP, oblong plate; PM, old pivoting muscle; QP, quadrate plate; Ra 1, first ramus; Ra 2, second ramus; TP, triangular plate. (Redrawn from Hermann and Chao, 1983.)

Retention of a Furcular Remnant in the Formicinae

Venom Evacuation and Directional Trajectory Manipulation

The venom sac and its duct in formicine ants have only sparse musculature (Hermann, 1983; Hermann and Chao, 1983). Such sparse musculature is not adequate for effecting rapid venom evacuation. Rapid venom evacuation is primarily controlled by abdominal hydrostatic pressure.

Changes in the direction of venom evacuation are due to: (1) intersegmental muscles along the abdominal wall; (2) dilator muscles that open the venom canal; (3) muscles inserting on a small sclerite that adheres to the venom canal (Figure 6.35); (4) muscles between the abdominal wall and the paired spiracular plates; and (5) muscles between the spiracular plates and the quadrate plates.

The small sclerite that receives muscles from the oblong plate is called a furcular remnant (Hermann, 1983; Hermann and Chao, 1983). Retention of this sclerite in a group of insects in which much of the venom apparatus has been lost points to the fact that the furcula or some homolog is found in all aculeates regardless of what has happened to the rest of the apparatus phyletically.

Review of Functions in Stinging Aculeates

In recapitulation, the venom duct in stinging aculeates enters the sting bulb dorsad of Dufour's gland. The duct adheres to the dorsal region of its inner wall. The furcula stands upright on the sting base, away from the venom duct in furculate species, effecting a rotating motion of the sting during the contraction of muscles inserting distad on the dorsal furcular apodeme. In nonfurculate species, these muscles insert on the anterior sting base extension, which lies adjacent to the venom duct. Their contraction causes a pivoting action of the sting.

Formicine Predecessor

Since the furcular remnant in formicine species adheres to the venom canal, similar to the way in which the sting of nonfurculate species and venom duct adhere, it was suggested by Hermann (1983) that formicine ants arose from nonfurculate ants, species that pivoted the sting rather than rotated it.

Abdominal muscles that facilitate the manipulative qualities of the venom duct (Figures 6.36 and 6.37) are intertergal extensors, intersternal contractors, and tergosternal dilators and retractors (Goyal, 1977). Muscles of the paired spiracular and quadrate plates function in moving the entire apparatus.

Sting pivoting is not as effective in sting manipulation as is sting rotation (Hermann and Chao, 1983). Since the furcula has been found in all of the aculeate groups and even in an internal parasitoid (Hermann and Morrison, 1980), I believe that all ants were furculate at one time; in those few ant species in which the pivoting muscles insert on the anterior sting base extension, the furcula has reunited with the sting base. Fusion indicates a reduced importance to the action of the sting as a defensive weapon.

Phyletic Sequence in Elaborational and Reductional Changes at the Sting Base

Since formicines have no sting but do possess a furcular homolog, it is likely that they have long ago traversed the phyletic path currently being independently followed by other extant formicids in the following sequence: (1) development of a furcula from the anterior portion of the sting base; (2) furcular elaboration in some species; (3) retention of the furcula but desclerotization of it and other parts of the venom apparatus (myrmicines); (4) a reuniting of

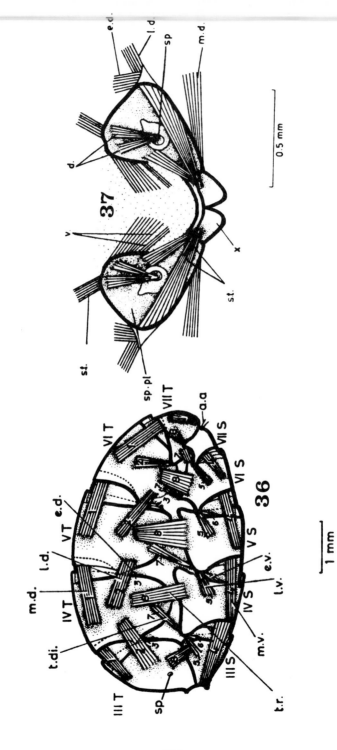

FIGURES 6.36 and 6.37. Abdominal plates of *Camponotus pensylvanicus* (redrawn from Goyal, 1977). (36) Mesal view of gastral wall, showing musculature; (37) dorsal view of spiracular plates (8th hemitergites), showing musculature.

the furcula to the sting base, with a retention of the old rotating muscles to pivot the sting (New World dorylines); (5) reduction of venom sclerites (Old World dorylines); (6) desclerotization of venom sclerites, especially at the sting tip (dolichoderines); and (7) reduction of first and second valvulae (lancets and sting, respectively), with further desclerotization but retention of the furcular remnant to effect venom evacuation (Formicinae). Venom dispersal is further facilitated in this latter group by particle-scattering hairs that line the acidopore.

Taxonomic and Phyletic Implications

Although taxonomic implications have not been discussed so far, there are some characters of the hymenopterous venom apparatus that are of taxonomic and phyletic importance (Hermann, 1983). It is believed by some investigators that the Aneuretinae, Dolichoderinae, and Formicinae have all arisen from nothomyrmeciines (Brown, 1954; Snelling, 1981). Also, *Nothomyrmecia* and *Myrmecia* are considered to be closely related (Brown, 1954; Kugler, 1980; Taylor, 1978). A well-developed furcula has been found in *Nothomyrmecia macrops* (Kugler, 1980) and this furcula has the general appearance and manner of articulation with the sting base that we find in the Ponerinae and Myrmeciinae. In particular, a lateral view of the venom apparatus shows that the furcula does not articulate to the top of the sting, but rather it articulates to the dorsolateral region of the sting base. The venom apparatus of *Aneuretus simoni* (Hermann, 1968b) is well developed, but there has been some desclerotization, and the apparatus lacks a furcula (fusion between the sting base and furcula has occurred). The dolichoderine venom apparatus (Pavan and Ronchetti, 1955; Hermann and Blum, 1981) is reduced from that of *Aneuretus*. Finally, the formicine venom apparatus is further reduced, and it has been implied in this chapter that this latter group has arisen from a nonfurculate predecessor, similar to the dolichoderines or aneuretines.

Further points of anatomical interest may be found in the structure of the gonostyli. Gonostyli with a stereotyped bilobed appearance have been found in the ants only in the subfamilies Myrmeciinae and Ponerinae thus far, two of the most primitive groups of formicids. Although Kugler (1978) described bilobity in the gonostyli of the myrmicine, *Myrmica emeryana*, and a thinning of the cuticle at a point on the distal half of the gonostylus of the myrmicines *Ephebomyrmex naegeli, Leptothorax longispinosus, Vollenhovia pedestris* and *Aphaenogaster rudis*, the general appearance of the bilobity is distinctly different from that in ponerines and myrmeciines. At this time I feel that this character of stereotyped gonostylar bilobity, consistently present in the Myrmeciinae and Ponerinae, has phyletic and taxonomic significance in those groups. Gonostyli in *Nothomyrmecia macrops* are unilobular, although Kugler (1980) noticed a cuticular thinning in the gonostyli of this species also. In this one character, *Nothomyrmecia* is more like ants other than the ponerines and myrmeciines. Sting morphology may support and does not refute this phyletic sequence from nothomyrmeciines to aneuretines, dolichoderines, and formicines. It would not support the rise of ponerines and other myrmeciines from *Nothomyrmecia* because of the derived condition of the gonostyli in *Nothomyrmecia macrops*. Combining the gonostylar and furcular characteristics strongly indicates that *Nothomyrmecia*, other myrmeciines, and ponerines all had a common ancestor.

Venom gland anatomy in these various groups (*Nothomyrmecia*, Dolichoderines, Formicines) does not support the final step in this sequence, that is, rise of the Formicinae from these other groups. The convoluted gland in the Formicinae, unlike in any other hymenopteran, is distinctly outside the reservoir (Blum and Hermann, 1978b; Wilson and Regnier, 1971). The convoluted gland in *Nothomyrmecia*, dolichoderines, and ant species in all subfamilies other than the Formicinae is distinctly within the reservoir. This anatomical difference between formicines and other ant subfamilies appears to be a very significant one. In addition, the venom constituents in formicine ants are quite different from those of all other ant species (Blum and Hermann, 1978a,b). Formic acid appears to be a diagnostic character for all species in the subfamily Formicinae, and although acidic constituents produced by a few species of ants in other subfamilies have been reported as formic acid, there is no strong evidence to indicate that this compound is indeed a constituent of the venoms of nonformicine ants (Blum and Hermann, 1978a). These very significant differences indicate that formicines probably separated from other ants before the development of these other groups, including the Nothomyrmeciinae.

REFERENCES

Alexander, R. D. 1974. The evolution of social behavior. *Annu. Rev. Ecol. Syst.* 325–379.
Blum, M. S., and H. R. Hermann. (1978a). Venoms and venom apparatuses of the Formicidae: Myrmeciinae, Ponerinae, Dorylinae, Pseudomyrme-

cinae, Myrmicinae, and Formicinae. In *Arthropod Venoms* (S. Bettini, ed.), *Handbook Experimental Pharmacology*, Vol. 48, pp. 801–869. Springer-Verlag, Berlin and New York.

Blum, M. S., and H. R. Hermann. (1978b). Venoms and venom apparatuses of the Formicidae: Dolichoderinae and Aneuretinae. In *Arthropod Venoms* (S. Bettini, ed.). *Handbook of Experimental Pharmacology*, Vol. 48, pp. 871–894. Springer-Verlag, Berlin and New York.

Brown, W. L. (1954). Remarks on the internal phylogeny and subfamily classification of the family Formicidae. *Insectes Sociaux* 1, 21–31.

Brown, W. L. (1975). Contributions toward a reclassification of the Formicidae. V. Ponerinae, tribes Platythyreini, Cerapachyini, Cylindromyrmecini, Acanthostichini and Aenictogitini. *Search Agric. Entomol.* (15), 5, 1–115.

Carpenter, F. M., and H. R. Hermann. (1979). Antiquity of Sociality in Insects. In *Social Insects* (H. R. Hermann, ed.), Vol. 2, pp. 81–89. Academic Press, New York.

Chadab-Crepet, R., and C. W. Rettenmeyer. (1982). Comparative behavior of social wasps when attacked by army ants or other predators and parasites. In *The Biology of Social Insects* (M. D. Breed, C. D. Michener, and H. E. Evans, eds.). Proc. 9th Inter. Cong. Stud. Soc. Insects, Boulder, Colorado.

D'Rozario, A. M. (1942). On the development and homologies of the genitalia and their ducts in Hymenoptera. *Trans. R. Entomol. Soc. London* 92, 362–415.

Eickwort, G. (1981). Presocial insects. In *Social Insects* (H. R. Hermann, ed.), pp. 199–280. Academic Press, New York.

Evans, H. O. (1958). The evolution of social life in wasps. *Proc. 20th Int. Congr. Entomol., Montreal, 1956,* 2, 449–457.

Flemming, H. O. (1957). Die Muskulatur und Innervierung des Wehrstachelapparates von Aculeaten. *Z. Morphol. Oekol. Tiere* 46, 321–341.

Gamboa, G. J. (1978). Intraspecific defense: Advantage of social cooperation among paper wasp foundresses. *Science* 199, 1463–1465.

Gibo, D. L. (1978). The selective advantage of foundress associations in *Polistes fuscatus*. A field study of the effects of predation on productivity. *Can. Entomol.* 110, 519–540.

Goyal, A. (1977). Studies on the skeleto-muscular system of the common Indian ant, *Camponotus compressus* Fabr. and bionomics of some ants from Delhi. Thesis, Univ. of Delhi, India.

Gupta, P. D. (1950). On the structure, development and homology of the female reproductive organs in Orthopteroid insects. *Ind. J. Entomol.* 10, 75–123.

Gustafson, J. F. (1950). The origin and evolution of the genitalia of the insecta. *Microentomology* 15, 35–67.

Hamilton, W. D. (1964). The genetical evolution of social behavior, I and II. *J. Theor. Biol.* 7, 1–52.

Hamilton, W. D. (1972). Altruism and related phenomena, mainly in social insects. *Annu. Rev. Ecol. Syst.* 3, 193–232.

Hanna, A. D. (1934). The male and female genitalia and the biology of *Euchalcidia caryobori* Hanna (Hymenoptera, Chalcidinae). *Trans. R. Entomol. Soc. London* 82, 107–136.

Hazeltine, W. E. (1967). Female genitalia of Hymenoptera and comparative morphology of male and female genital segments of Bombinae (Hymenoptera, Apidae). *Indiana Agric. Exp. Stn. Res. Bull.* 833, 1–36.

Hermann, H. R. (1968a). The hymenopterous poison apparatus. IV. *Dasymutilla occidentalis* (Hymentopera: Mutillidae). *J. Ga. Entomol. Soc.* 3, 1–10.

Hermann, H. R. (1968b). The hymenopterous poison apparatus. V. *Aneuretus simoni. Ann. Entomol. Soc. Am.* 61, 1315–1317.

Hermann, H. R. (1968c). The hymenopterous poison apparatus. VIII. *Simopelta oculata* (Hymenoptera: Formicidae: Ponerinae). *J. Ga. Entomol. Soc.* 3, 163–166.

Hermann, H. R. (1969). The hymenopterous poison apparatus: Evolutionary trends in three closely related subfamilies of ants (Hymenoptera: Formicidae). *J. Ga. Entomol. Soc.* 4, 123–141.

Hermann, H. R. (1971). Sting autotomy, a defensive mechanism in certain social Hymenoptera. *Insectes Sociaux* 18, 111–120.

Hermann, H. R. (1979). Insect sociality—An introduction. In *Social Insects*, (H. R. Hermann, ed.), Vol. 1, pp. 1–33. Academic Press, New York.

Hermann, H. R. (1983). Directional trajectory manipulation of venom in formicine ants, *Sociobiology* 8: 105–117.

Hermann, H. R., and M. S. Blum. (1981). Defensive mechanisms in the social Hymenoptera. In *Social Insects* (H. R. Hermann, ed.), Vol. 2, pp. 77–197, Academic Press, New York.

Hermann, H. R., M. S. Blum, J. W. Wheeler, W. L. Overal, J. O. Schmidt, and J. T. Chao (1984). Comparative anatomy and chemistry of the venom apparatus and mandibular glands in *Dinoponera grandis* (Guerin) and *Paraponera clavata* (Fabr.) (Hymenoptera: Formicidae: Ponerinae). *Ann. Entomol. Soc. Am.* (in press).

Hermann, H. R., and J. T. Chao. (1983). Furcula, a major component in the hymenopterous venom apparatus. *Inter. J. Ins. Morph. Embryol.* 12, 321–377.

Hermann, H. R., and M. Douglas. (1976a). Sensory structures on the venom apparatus of a primitive ant species. *Ann. Entomol. Soc. Am.* 69, 681–686.

Hermann, H. R., and M. Douglas. (1976b). Comparative survey of the sensory structures on the sting and ovipositor of hymenopterous insects. *J. Ga. Entomol. Soc.* 11, 223–239.

Hermann, H. R., and J. W. Krispyn. (1975). The hymenopterous venom apparatus. XIV. *Vespula maculata* (Hymenoptera: Vespidae). *J. Ga. Entomol. Soc.* 10, 307–313.

Hermann, H. R., and G. Morrison. (1980). Ovipositor and associated glands of *Microplitis croceipes* (Braconidae). *J. Ga. Entomol. Soc.* 15, 479–485.

Hunt, A. N., and H. R. Hermann. (1970). The hymenopterous poison apparatus. X. *Polistes annularis* (Hymenoptera: Vespidae). *J. Ga. Entomol. Soc.* 5, 210–216.

Jeanne, R. L. (1979). Construction and utilization of multiple combs in *Polistes canadensis* in relation to the biology of a predaceous moth. *Behav. Ecol. Sociobiol.* 4, 293–310.

Jeanne, R. L. (1982). Predation, defense, and colony size and cycle in the social wasps. In *The Biology of Social Insects* (M. D. Breed, C. D. Michener, and H. E. Evans, eds.), pp. 280–284. Westview Press, Boulder, Colo.

Kugler, C. (1978). A comparative study of the myrmicine sting apparatus (Hymenoptera, Formicidae). *Stud. Entomol.* 20, 413–548.

Kugler, C. (1980). The sting apparatus in the primitive ants *Nothomyrmecia* and *Myrmecia*. *J. Aust. Entomol. Soc.* 19, 263–267.

Lamborn, W. A. (1915). Second report on *Glossina* investigations in Nyasa-land. *Bull. Entomol. Res.* 6, 252–255.

Lin, N., and C. D. Michener. (1972). Evolution of sociality in insects. *Q. Rev. Biol.* 47, 131–159.

Malyshev, S. I. (1968). *Genesis of the Hymenoptera and the Phases of Their Evolution.* Methuen, London.

Maschwitz, U. (1964). Gefahrenalarmstoffe und Gefahrenalarmierung bei sozialen Hymenopteron. *Z. Vergl. Physiol.* 47, 596–655.

Maschwitz, U., and W. Kloft. (1971). Morphology and function of the venom apparatus of insects, bees, wasps, ants and caterpillars. In *Venomous Animals and Their Venoms* (W. Bucherl and E. E. Buckley, eds.), Vol. 3 (44), pp. 31–60. Academic Press, New York.

Matsuda, R. (1957). Comparative morphology of the abdomen of a machilid and rhaphidiid. *Trans. Am. Entomol. Soc.* 83, 39–63.

Matsuda, R. (1958). On the origin of the external genitalia of insects. *Ann. Entomol. Soc. Am.* 51, 84–94.

Matsuda, R. (1976). *Morphology and Evolution of the Insect Abdomen.* Pergamon Press, New York.

Maynard Smith, J. (1978). The evolution of behavior. In *Evolution*, pp. 92–101. W. H. Freeman & Co., San Francisco.

Michener, C. D. (1944). A comparative study of the appendages of the eighth and ninth abdominal segments of insects. *Ann. Entomol. Soc. Am.* 37, 336–351.

Oeser, R. (1961). Vergleichend-morphologische Untersuchungen über den Ovipositor der Hymenopteron. *Mitt. Zool. Mus. Berl.* 37, 1–119.

Owen, M. D., and A. R. Bridges. (1976). Aging in the venom glands of queen and worker honey bees (*Apis mellifera* L.): Some morphological and chemical observations. *Toxicon* 14, 1–5.

Pavan, M., and G. Ronchetti. (1955). Studi sulla morfologia esterna e anatomia interna dell operaia di *Iridomyrmex humilis* Mayr e ricerche chimiche e biologiche sulla iridomercina. *Atti Soc. Ital. Sci. Nat. Mus. Civ. Stor. Nat. (Milan)* 94, 379–477.

Robertson, P. L. (1968). A morphological and functional study of the venom apparatus in representatives of some major groups of Hymenoptera. *Aust. J. Zool.* 16, 133–166.

Scudder, G. G. E. (1957a). Reinterpretation of some basal structures in the insect ovipositor. *Nature (London)* 180, 340–341.

Scudder, G. G. E. (1957b). The systematic position of *Dicranocephalus* Hahn. 1826 and its allies (Hemiptera-Heteroptera). *Proc. R. Entomol. Soc. London,* (*Ser. A*) 32, 147–158.

Scudder, G. G. E. (1959). The female genitalia of the Heteroptera: Morphology and its bearing on classification. *Trans. R. Entomol. Soc. London* 111, 405–467.

Scudder, G. G. E. (1961a). The comparative morphology and interpretation of the insect ovipositor. *Trans. R. Entomol. Soc. London* 113, 25–40.

Scudder, G. G. E. (1961b). The functional morphology and interpretation of the insect ovipositor. *Can. Entomol.* 93, 267–272.

Scudder, G. G. E. (1964). Further problems in the interpretation and homology of the insect ovipositor. *Can. Entomol.* 96, 405–417.

Seeley, T. D. (1982). Colony defense strategies of honeybees in Thailand. In *The Biology of Social Insects* (M. D. Breed, C. D. Michener, and H. E. Evans, eds.). Westview Press, Boulder, Colo.

Sharov, A. G. (1966). *Basic Arthropodan Stock.* Pergamon Press, New York.

Smith, E. L. (1969). Evolutionary morphology of the external insect genitalia. I. Origins and relationships to other appendages. *Ann. Entomol. Soc. Am.* 62, 1051–1079.

Smith, E. L. (1970). Evolutionary morphology of the external insect genitalia. 2. Hymenoptera. *Ann. Entomol. Soc. Am.* 63, 1–27.

Snelling, R. R. (1981). Systematics of social Hymenoptera. In *Social Insects* (H. R. Hermann, ed.) Vol. 2, pp. 369–453, Academic Press, New York.

Snodgrass, R. E. (1931). Morphology of the insect abdomen. Part I. General structure of the abdomen and its appendages. *Smithson. Misc. Collect.* 85, 1–128.

Snodgrass, R. E. (1933). Morphology of the insect abdomen. Part II. The genital ducts and the ovipositor. *Smithson. Misc. Collect.* 89, 1–148.

Snodgrass, R. E. (1935). *Principles of Insect Morphology.* McGraw-Hill, New York.

Spradbery, J. P. (1973). *Wasps.* Univ. of Washington Press, Seattle.

Starr, C. K. (1981). Defensive tactics of social wasps. Doctoral Dissertation, Univ. of Georgia, Athens, 108 pp.

Steiner, A. L. (1975). Description of the territorial behavior of *Podalonia valida* (Hymenoptera, Sphecidae) females in Southeast Arizona, with remarks on digger wasp territorial behavior. *Quaest. Entomol.* 11, 113–127.

Steiner, A. L. (1978). Observations on spacing, aggressive and lekking behavior of digger wasp males of *Eucerceris flavocincta* (Hymenoptera: Sphecidae: Cercerini). *J. Kan. Entomol. Soc.* 51, 492–498.

Stumper, R. (1960). Die Giftsekretion der Ameisen. *Naturwissenschaften* 47, 457–463.

Taylor, R. W. (1978). *Nothomyrmecia macrops*: A living-fossil ant rediscovered. *Science* 201, 979–985.

Trivers, R. L. (1971). The evolution of reciprocal altruism. *Q. Rev. Biol.* 46, 35–57.

Trivers, R. L. and H. Hare. (1976). Haplodiploidy and the evolution of social insects. *Science* 191, 249–263.

Trojan, E. (1935). Zur Frage der Oligomerie weiblicher Akuleaten. *Z. Morphol. Oekol. Tiere* 30, 597–628.

Weiss, J. (1978). Vergleichende Morphologie des Stachelapparates bei den Vier *Apis*-Arten (Hymenoptera: Apidae). *Apidologie* 9, 19–32.

Wilson, E. O. (1971). *The Insect Societies.* Belknap Press, Cambridge, Mass.

Wilson, E. O., and F. E. Regnier. (1971). The evolution of the alarm-defense system in the formicine ants. *Am. Nat.* 105, 279–289.

Zander, E. (1899). Beitrage zur Morphologie des Stacheapparates der Hymenopteran. *Z. Wiss. Zool.* 66, 289–333.

Zander, E. (1911). *Handbuch der Bienenkunde in ein Zeldarstellungen. Der Bau der Biene*, Vol. 3. Ulmer, Stuttgart, Federal Republic of Germany.

GLOSSARY

aculeate—This is a term reserved for those hymenopterous insects that possess a sting or have arisen from individuals that possess a sting rather than an ovipositor, the egg passing out of the gonopore just anterior to the gonocoxapophyseal base. All aculeates do not sting; many species have venom apparatuses that are reduced to the point that stinging is impaired or impossible.

altruistic behavior—Behavior that benefits another oganism while being apparently detrimental to the organism performing the behavior, benefit and detriment being defined in terms of contribution to inclusive fitness (Trivers, 1971).

anterior extension of sting base—Hymenopterous insects that have this extension are either prefurculate (Symphyta) or postfurculate (Dorylinae, Cerpachyinae, Dolichoderinae, Aneuretinae). In both cases, muscles that insert on the furcula in furculate species (most aculeates) insert in these species on this anterior extension.

anterior gonocoxapophyseal muscles—These paired muscles originate on each of the second rami and insert on the anterior region of the sting base in both furculate and nonfurculate Hymenoptera. They function in sting levation. They are antagonistic to the portion of the posterior gonocoxapophyseal muscles that function in sting depression.

barbs—Barbs are found on the distal lancet tip in many aculeates. They are also found on the distal sting tip in most species of *Apis*. Enlarged barbs function in sting autotomy.

convoluted gland—Portion of the venom-producing structures in certain aculeates. It is not found in bees but is homologous in those insects to the proximal region of the filamentous gland. In most aculeates it has invaginated into the venom reservoir. In formicine ants it is outside of the latter structure.

defense (group)—Defensive behavior in which a number of individuals warn together or leave the nest together in an attack. Such defensive behavior is found only in social insects.

defense (individual)—Defensive behavior exhibited by a solitary species or by a social species that is away from its nest. It is usually not demonstrated unless the insect's life is in jeopardy.

defensive behavior—Any behavior used by an insect or group of insects to defend its life, its coinhabitants, or its colony.

desclerotization—A reduction process having to do with either a lightening of the color of sclerites or their actual reduction in size.

Dufour's gland—The ventral-most gland that enters the sting base in aculeate Hymenoptera. It is most often a simple, elongate sac surrounded by secretory cells, but it may be globate or bilobed in certain social species. It has erroneously been called alkaline gland by many investigators.

elaboration—Change toward a more complex structure or a greater number of parts.

escape—A maneuver in which an individual avoids an encounter with an intruder by leaving. Escape is preferred by solitary wasps and social species away from their nest.

eusocial—A term that pertains to the highest social position according to the definition by Wilson (1971).

exsertile—Refers to the sting when it protrudes from the posterior tip of the abdomen.

fulcral arms—Slender-paired structures that may have originated from the ninth sternum and extend dorsad from the ventral region of the anterior sting base. Their ventral points of articulation with the sting base function as a transverse horizontal axis around which the functions of sting depression and levation occur.

fulcral points—There are three such points of bending, usually discussed in the venom apparatus. A fulcral point at the base of the fulcral arms functions in sting depression and levation. A fulcral point at the point of articulation between furcula and sting base functions in moving the furcula anteriad and posteriad, effecting movements of levation and depression. A fulcral point at the gonostylar base affects sting manipulation (usually bilateral movement) during the contraction of furcular muscles.

furcula—A small sclerite that has evolved from the anterior sting base. It functions in sting depression, levation, and rotation. The furcula exists in most aculeates but has been reduced in some ant species and stingless bees.

furcular remnant—A small sclerite in formicine ants that has arisen from the furcula of stinging aculeates.

furculate—A term applied to hymenopterans that have a furcula. Those hymenopterans that represent a form prior to the formation of the furcula or following fusion between the furcula and sting base are considered to be nonfurculate.

gonangulum—Often referred to as the triangular plates, these paired structures, according to Scudder (1957a,b, 1959, 1961a,b, 1964), have arisen from precoxal plates of the ninth segment.

gonapophyses—Extensions from the eighth and ninth (genital) abdominal segments in ovipositor-bearing insects.

gonostyli—Extensions from the paired oblong plates. Referred to by Scudder as gonoplacs, these structures are sensory in nature and also form a sheath over the ovipositor or sting. They are often considered to be the third valvulae that have originated from definitive styli, but Matsuda (1976) stated that they, in fact, may not represent true styli.

hemitergites—Paired plates of the eighth and ninth abdominal terga (spiracular and quadrate plates, respectively). The eighth hemitergites are usually joined mesally by a posterodorsal bar while the ninth hemitergites have no sclerotized connection.

internalization—A developmental process in which the posteriormost abdominal segments move into the seventh segment. This process is actually a two-stage one in which the eighth and following segments move into the seventh and later the ninth and following segments move into the eighth. This process is known also as interiorization (Matsuda, 1976).

intruder—An individual that approaches the nest of a social hymenopteran species, whether the approach is planned or accidental. Such intrusion often provokes the species intruded on to defend its nest, the queen, or the immatures that the nest houses.

lancets—The paired first valvulae. Lancets often support a valve on their anterodorsal border and barbs on their distal end. They extend anteriad by way of the first rami and articulate to the triangular plate on the latter structure's anterior surface.

nonfurculate—Any species that does not have a furcula, whether it is prefurculate (e.g., sawfly or external parasitoid) or postfurculate (e.g., doryline and dolichoderine ants).

Nothomyrmecia macrops—The most primitive extant ant species, collected most recently in Australia by Taylor (1978).

oblong plates—The paired second valvifers of the ninth abdominal segment. They articulate with the sting base ventrally by way of the second rami and posteriorly with the gonostyli (third valvulae).

occlusor muscle—A closing muscle, used here in describing muscles that pass over the venom duct to close it when venom is not being evacuated from the venom reservoir.

parasitoid—The lower Apocrita. These insects deposit their eggs on or in the host so that their larvae hatch and devour the host.

posterior gonocoxapophyseal muscles—These paired muscles originate on the second valvifers (oblong plates) and insert either on the ovipositor or sting base or on the lateral region of the furcula.

predator—An organism considered to be free living and that catches another organism and devours it. Most solitary and social aculeates (upper Apocrita) that catch prey are considered to be predatory, while members of the lower Apocrita (Terebrantia) are considered parasitoids.

quadrate plates—Paired ninth hemitergites which have moved, along with most of the venom apparatus, into the eighth segment through a second internalization.

rami—Paired, slender structures linking valvulae and valvifers of the eighth and ninth segments. The first rami link the lancets to the triangular plates and the second rami link the sting to the oblong plates.

reduction—A process whereby parts are fused or lost.

sclerotization—A process whereby sclerites either become darker in color or harder or a combination of the two.

spiracular plates—Paired eighth hemitergites, joined dorsomesally by a thin bar.

sting—A structure made up of the fused second valvulae dorsally and laterally and the paired lancets ventrally.

sting ascendance—An upcurving of the sting. The terms ascendance and descendance have been used here to replace recurved and decurved because they are more accurate in their meaning. Although decurved implies a downward curve, the term recurved means curved backward or inward.

sting autotomy (= Autothysis)—A process whereby lancet barbs and sometimes barbs on the distal sting tip are enlarged so that they cause the sting to pull away from the stinging insect to remain at the sting locus.

sting bulb—The enlarged anterior sting base into which the venom duct and Dufour's gland exit. The dorsal region of the venom duct adheres to the anterodorsal region of the sting bulb.

sting descendance—A downward curving of the sting, found well developed in vespoids and best demonstrated in the Mutillidae.

sting depression—Lowering of the distal sting tip. Depression comes about with a contraction of muscles that insert on the ventrolateral portion of the furcula or the anterior sting or ovipositor base extension.

sting levation—An antagonistic movement to sting depression, effected by a contraction of muscles originating on the second rami and inserting on the anterior sting or ovipositor base.

sting pivoting—A bilateral movement of the sting caused by a contraction of muscles originating on the oblong plates and moving mesad to an insertion point on the anterior sting base.

sting recurvity—An upward curving of the sting, found to be very well developed in the ant species *Onychomyrmex hedleyi* (Ponerinae) (described here in terms of sting ascendance).

sting rotation—A circular movement of the sting in furculate species, caused by a contraction of muscles originating on the paired oblong plates and moving mesad to an insertion point on the laterodistal region of the furcula.

sting sheath—A cover over the sting formed by the two lateral halves of the third valvulae.

triangular plates—Paired plates that articulate to the lancets through the first rami, with the quadrate plates by a dorsal apodeme and weakly with the oblong plates by a ventral apodeme.

trichoid sensilla—Sensory hairs found in several locations on the venom apparatus, for example the hair plates on the oblong plates, the gonostyli, and the second rami.

venom apparatus—The apparatus composed of sclerites of the eighth and ninth abdominal segments in female aculeates and the associated glands.

venom duct—Duct from the venom reservoir to the sting base.

venom reservoir—The sac that stores venom after it is synthesized in the filamentous and convoluted glands.

Index

abdomen, pumping of, 15

abdominal bursting, as defensive mechanism, 11

acacia, symbiosis with *Pseudomyrmex ferruginea,* 129

Acanthognathus, mandibular clicking and, 13

Acanthotermes acanthothorax, frontal weapon, 182

acoustic signals, in vespines, 74–75

aculeate Hymenoptera, venom apparatus in, 201–243; defensive behaviors increase in, 226–227; description of, 203–204; elaboration, 214–227; furcular remnant retention in Formicinae, 234–238; fusion between sting base and furcula, 229–231; loss of sting function, 231; origin and significance of furcula, 209–214; reduction, 227–243; sting coiling in Mutillidae, 221–226; venom sclerites, 204–209

aggression: and juvenile hormone, 19; physiology of, 18–20

aggression types, 18

Akre, R. D., 59

alarm pheromones, 15; amount, 126; body postures and, 126; nest mates reactions to, 126; sources of, 125; in vespines, 75

alert posture, 12

Allee, W. C., 2

altruism, and defensive mechanisms, 3, 203

Amblyopone australis, vibrational alarm behavior, 124

Amitermes evuncifer: frontal gland secretions, 183; frontal weapon, 183

Ancistocercus inficitus, 39

Andrenidae, 7

Aneuretinae, furcula fusion, 229

Aneuretus simoni: sting function loss, 231; unilobular gonostyli, 232

animal aggregations, 2; lengthened survival, 2

animal groups, defensive behavior, 2

animal society, 1

Anochetus, mandibular clicking and, 13

Anoplura, ovipositor loss, 203

ant(s): alarm behavior of, 123–128; alarm pheromones in, 113; chemical alarm defense system in, 13, 125; chemical secretions of, 100; chemical territorial demarcation, 131; chemical weapons, 118–123; defense against other ants, 99, 107; defensive behavior in, 95–150; defensive behavior modification, 226; defensive glands in, 118–123; food resources, defense of, 128–129; group defense, 123–128; interactions among themselves, 99; intraspecific territorial competition, 131–132; modified fighting behavior, 137; morphological structures of, 100–103; nest construction, 105; nest plugging, 103–104; poison gland of, 111; as predators, 100; protection against weather, 104; pro-

Rau, P., 39
Reed, H. C., 59
repellent chemicals, defensive
    smearing, 14
reproductive reproduction, 40–41
Reticulitermes, frontal gland,
    167–169; cuticle of, 195; his-
    tology of, 169
retreat, of polistine wasps, 16
Rhinotermitidae, 164–177; frontal
    gland, 173–177
Richards, O. W., 33, 39
rodents, and yellowjacket colonies,
    77
Roseler, P. F., 19
Rostrotermes, 191
"rubbing," 44

salivary glands: cellular structure,
    157; defensive function, 155,
    192; neurosecretory innerva-
    tion, 161; reservoirs, 157; secre-
    tions, 161
sawflies, 210
Schedorhinotermes, frontal gland,
    173–175; cuticle of, 195; his-
    tology of, 174–175; secretions
    of, 174
Scheven, J., 48
Scudder, G. G. E., 205
Serritermitidae, 5
sesquiterpenes, 183
Sharov, A. G., 205
Smith, J. M., 3
social insects: altruism and, 3; com-
    munication system, 15; defen-
    sive strategies, 2; insect brood
    parasitoids and, 2; insect preda-
    tors and, 2; natural enemies, 2;
    parasites and, 2; territorialism,
    16–18; vertebrate predators
    and, 2; warning behavior, 9–10
social parasitism, in *Vespula*, 83
soldier termites: cibarial glands of,
    152–154; epidermal glands of,

152–155; frontal gland of, 163;
    salivary glands of, 155–161;
    salivary reservoirs, 157
sound production, by honey bee
    colonies, 74
Sphaerotermes sphaerothorax, frontal
    pore, 179
Sphecidae, 5
Sphecophaga, as vespine parasite,
    86–88
Spradbery, J. P., 61
Starr, C. K., 42
Steiner, A. L., 202
sticky fluids, use of, 16
sting: depressor muscles, 208, 213;
    primary region, 208
sting autotomy, 220, 226; in ants,
    112; lancet barbs and, 226
sting base, 213; furcula fusion and,
    229–231
sting bulb, 208
sting coiling, 221–226; furcular
    elongation, 224
sting disfunction, 231–234; in
    Myrmicinae, 227
sting furcula: evolution of, 210;
    functions of, 209–210; origin
    and significance of, 209–214;
    shape of, 209; sting depression
    and, 209
stinging apparatus: functional
    morphology, 114–117; pumping
    mechanism, 115; reduction,
    114–117, 132–136; surface
    venoms production, 115; trail
    pheromones production,
    114–115
sting levation, 209
sting sclerites, reduction, 227
sting shaft elaboration, 220
stridulation, 16
Strumigenys, mandibular clicking and,
    13
swarming, 45–46
symbiosis, in ants, 129
Symphyta, ovipositor, 210
Syntermes soldiers: frontal gland of,
    187; mandibles, 187

ABOUT THE EDITOR

Dr. Henry R. Hermann is an Associate Professor of Entomology at the University of Georgia where he has held a research and teaching position since 1967. Prior to college, Dr. Hermann was photographer and editor of two Air Force newspapers. He received his B.S. degree from New Orleans University in 1963, and M.S. and Ph.D. degrees from Louisiana State University in Baton Rouge, Louisiana, in 1965 and 1967.

He is former Associate Editor of the *Journal of the Georgia Entomological Society* and has recently edited a four-volume series, "Social Insects," that was published by Academic Press from 1979 to 1982. In addition, Dr. Hermann currently has over 70 scientific publications primarily on the morphology and behavior of defensive mechanisms in the social Hymenoptera.

Dr. Hermann's research has taken him to countries in North, Central, and South America. Of particular interest to him in tropical locations are some of the primitive ants and social wasps.

He has taught courses in entomology at every level since the commencement of his appointment at the University of Georgia, including undergraduate, graduate, and honors courses. During his teaching career, he has been awarded Teacher of the Year honors, being recognized at both the departmental and university level.